PROCEEDINGS

OF

THE JOHNS HOPKINS WORKSHOP

ON

CURRENT PROBLEMS IN PARTICLE THEORY

10

PREVIOUS JOHNS HOPKINS WORKSHOP PROCEEDINGS

1	BALTIMORE	'74	. . . .
2	BALTIMORE	'78	. . . .
3	FLORENCE	'79	. . . .
4	BONN	'80	Lattice Gauge Theories. Integrable Systems.
5	BALTIMORE	'81	Unified Field Theories and Beyond.
6	FLORENCE	'82	Lattice Gauge Theories. Supersymmetry and Grand Unification.
7	BONN	'83	Lattice Gauge Theories. Supersymmetry and Grand Unification.
8	BALTIMORE	'84	Particles and Gravity.
9	FLORENCE	'85	New Trends in Particle Theory.

The Johns Hopkins Workshop on Current Problems in Particle Theory are organized by the following universities:

UNIVERSITY OF BONN

UNIVERSITY OF FLORENCE

THE JOHNS HOPKINS UNIVERSITY

UNIVERSITY OF LANZHOU

ORGANIZING COMMITTEE

Gabor DOMOKOS (Johns Hopkins)
Susan KOVESI-DOMOKOS (Johns Hopkins)
Roberto CASALBUONI (Florence)
Luca LUSANNA (Florence)
Klaus DIETZ (Bonn)
Vladimir RITTENBERG (Bonn)
Yi-Shi DUAN (Lanzhou)
Gong-Ou XU (Lanzhou)

PROCEEDINGS OF THE JOHNS HOPKINS WORKSHOP ON CURRENT PROBLEMS IN PARTICLE THEORY 10

Bonn, 1986
September 1—3)

INFINITE LIE ALGEBRAS AND CONFORMAL INVARIANCE IN CONDENSED MATTER AND PARTICLE PHYSICS

Edited by
K. Dietz and V. Rittenberg

World Scientific

Published by

World Scientific Publishing Co Pte Ltd.
P. O. Box 128, Farrer Road, Singapore 9128

Library of Congress Cataloging-in-Publication data is available.

PROCEEDINGS OF THE 10TH JOHNS HOPKINS WORKSHOP ON CURRENT PROBLEMS IN PARTICLE THEORY – INFINITE LIE ALGEBRAS & CONFORMAL INVARIANCE IN CONDENSED MATTER & PARTICLE PHYSICS

ISBN 9971-50-240-2

Printed in Singapore by Kim Hup Lee Printing Co. Pte. Ltd.

FOREWORD

This is the tenth in a series of Workshops on Current Problems in Particle Theory. Since in the last four years the research done in condensed matter (phase transitions in two-dimensional systems) and particle physics (string theory) has been making use of the same infinite Lie algebras, the Organizing Committee decided to bring together people working in both fields. At the same time we thought it useful to invite mathematicians who did relevant work for physical applications. Last but not least we thought about experimental verifications but those are, unfortunately, for the time being, only in solid state physics.

The Workshop was based on invited talks which are all contained in these Proceedings, except for those of F. Englert, W. Nahm and S. Shenker who for various understandable reasons could not send in their manuscripts in due time.

We would like to thank the Deutsche Forschungsgemeinschaft and the Johns Hopkins University for financial support. We also thank Mrs. D. Faßbender and Mrs. M. Künkel for their dedicated help with the organization and editorial work connected with the Workshop.

The Organizing Committee

CONTENTS

PROCEEDINGS

OF

THE JOHNS HOPKINS WORKSHOP

ON

CURRENT PROBLEMS IN PARTICLE THEORY

10

QUANTUM SPIN CHAINS: EXPERIMENTAL REALIZATIONS OF KAC-MOODY ALGEBRAS

Ian Affleck*
Joseph Henry Laboratories
Princeton University
Princeton, New Jersey 08544, U.S.A.

ABSTRACT

A brief review of recent work on the application of Kac-Moody algebras to the study of the critical behaviour of quantum spin chains is given.

*Supported in part by National Science Foundation Grant # PHY80-19754and by the A.P. Sloan Foundation.

Conformal field theory can equally well be applied to classical two-dimensional systems near their critical temperatures and to critical one-dimensional quantum systems near zero temperature. (For more details see Ref. (1) and references therein.) By a critical one-dimensional quantum system I mean a system which has low energy excitiations obeying a linear dispersion relation $E = v|q|$. Here v plays the role of the velocity of light in the corresponding (1+1) dimensional quantum field theory. In such systems the Hamiltonian is expected to consist of a Lorentz covariant part plus irrelevant operators. Furthermore, the Lorentz covariant part should consist of a conformally invariant part plus irrelevant operators. These additional operators are only important in the higher energy part of the Hilbert Space.

An example is provided by a one-dimensional system of electrons. These may be mobile or localized on atoms, as in an antiferromagnet. In either case the system generally has symmetry under rotations of the electron spin direction. In some cases the full SU(2) symmetry is conserved, in others only a U(1) subgroup. Thus we expect the corresponding conformal field theory to have conserved currents. But in a conformal field theory conserved currents alway come in chiral pairs, $J_L(x_-)$, $J_R(x_+)$ where $x_\pm$ are light-cone co-ordinates. In the SU(2) case, the most general commutation relations consistent with conformal invariance is the Kac-Moody algebra:

$$[J_L{}^a(x_-),J_L{}^b(y_-)] = i\varepsilon^{abc}J_L{}^c\delta(x_--y_-) + (k/2\pi)\,\delta^{ab}\delta'(x_--y_-).$$

This algebra only has unitary representations if the central charge, k, is an integer. If we further assume that the energy-momentum tensor is quadratic in these currents, an assumption that arises naturally from bosonization of a fermionic system or from a spin Hamiltonian, then there is a unique theory for each positive integer value of k, the Wess-Zumino-Witten non-linear σ-model (WZW model)[2]. Conformal plus Kac-Moody symmetry completely determine all properties of these theories[2-5].

The specific heat of the system at low temperatures is determined by the linear part of the dispersion curve and has the form $C=\pi Tc/3v$ where c is a dimensionless number. c can be shown[6] to be the conformal anomaly parameter, which determines the strength of the two-point function of the energy-momentum tensor. Thus c can be interpreted as the number of degrees of freedom. It is one for a free boson or (Dirac) fermion. In general it can be fractional for an interacting system.

Similarly, the low-temperature susceptibility for a conserved total spin operator is determined only by the critical excitiations and takes the form $\chi = k/2\pi v$ where k is the strength of the two-point function of the conserved current. It is determined by the Schwinger term in the current commutator, and in the SU(2) case, is the Kac-Moody central charge.

Finally the anomalous dimensions of operators determine the correlation functions at sufficiently low frequencies, momenta and temperatures. At zero-temperature a primary field of dimension $\eta/2$ has the correlation function

$$G(x,t) = (x-t-i\varepsilon)^{-\eta/2}(x+t+i\varepsilon)^{-\eta/2}.$$

The finite-T correlation function can be found by making a conformal transformation from the infinite plane to the cylinder of circumference 1/T, to give the imaginary time Green's function[7] and then making the analytic continuation back to real time. This simply makes the replacement $x_\pm \pm i\varepsilon \rightarrow [\sinh \pi T(x_\pm \pm i\varepsilon)]/\pi T$ in the above formula. The Fourier transform of the static correlation function can be measured from the quasi-static neutron scattering cross-section. It has an approximately Lorentzian peak of width $\pi T\eta$, and height $\propto T^{-(1-\eta)}$ and a tail $\propto q^{-(1-\eta)}$. The Fourier transform of the time-dependent Green's function determines the inelastic neutron scattering cross-section. At fixed momentum there is power-law behaviour at T=0 :

$$G(E,q) \propto (E-|q|)^{-(1-\eta/2)} \quad (E>|q|)$$

with G=0 for $E < |q|$. At finite T, there is a peak at $E \approx |q|$ of width $\pi T\eta$ and a second peak at $E \approx -|q|$ which is suppressed by a Boltzman factor.

For the WZW models all the critical numbers are known exactly[3]. $c=3k/(2+k)$ and, for the lowest dimension primary field, $\eta=3/(2+k)$. To proceed, we need to identify various spin Hamiltonians with a particular value of k and find an operator mapping onto the field theory. Following the pioneering work of Luther and Peschel[8], we proceed by expressing the spins in terms of lattice fermions, passing to the continuum limit and then bosonizing.

Beginning with the Hamiltonian

$$H = \Sigma \mathbf{S}_i \cdot \mathbf{S}_{i+1}, \quad \mathbf{S}^2 = s(s+1)$$

we may represent the spins by fermions with 2 spin states and $n_c=2s$ colour states:

$$\mathbf{S} = (1/2)\psi^{+i}\sigma\psi_i \ (i = 1,2,3,\ldots n_c)$$

We must enforce the constraint that there are n_c fermions in a colour singlet state. This is totally antisymmetric in colour and hence, by fermi statistics, totally symmetric in spin, thus defining spin $s=n_c/2$. Note that half the available fermion levels are filled. We first consider the free fermion system with these quantum numbers and a half-filled band and put in the constraints later. The free Hamiltonian is

$$H = -\Sigma[\psi^+{}_i\psi_{i+1} + \psi^+{}_{i+1}\psi_i]$$

The dispersion relation is $E = v\cos qa$ (where a is the lattice spacing). Since the band is half-filled, low-energy excitations only involve the states near the two branches of the Fermi surface, $q \approx \pm\pi/2a$. Thus we pass to the continuum limit by keeping only Fourier modes of ψ close to these two values:

$$\psi(x) \approx i^{x/a}\psi_L(x) + (-i)^{x/a}\psi_R(x)$$

where ψ_L and ψ_R are slowly varying fields. H becomes the Hamiltonian for free Dirac fermions. The left-handed current in this free theory is $\mathbf{J}_L = (1/2)\psi_L{}^{+i}\sigma\psi_{Li}$ and it obeys the Kac-Moody algebra with $k=n_c$, since each colour makes an equal contribution to the Schwinger term. The conformal anomaly parameter is $c=2n_c$, the number of Dirac fermions. To obtain the critical theory for the spin chain, we must account for the constraint. Its role is essentially to produce a gap for excitations of colour or charge (fermion number). To find the critical theory we must decouple the colour and charge degrees of freedom leaving only the massless spin excitations. This can be accomplished by a form of non-abelian bosonization[2,1]. The free theory is equivalent to three decoupled massless theory, an SU(2) WZW model with $k=n_c$, representing the spin degrees of freedom, an SU(n_c) WZW model with k=2 representing the colour degrees of freedom and a free boson representing the charge degrees of freedom. It can be shown that the free fermion energy-momentum tensor can be expressed quadratically in these three sets of currents, and the same form is obtained from the bosonized theory. The values of c for the three parts of the theory are $3n_c/(2+n_c)$, $2(n_c^2-1)/(2+n_c)$ and 1. These add up to $2n_c$. The commutation relations and Green's functions of the currents in the bosonic theories are those of the free fermions. In particular the spin currents obey the Kac-Moody algebra with $k=n_c$. It thus appears very plausible that the effect of the constraints is simply to decouple the colour and charge sectors leaving the SU(2) $k=n_c$ WZW model as the critical theory for the $s=n_c/2$ spin chain. This then leads to predictions of the susceptibility and specific heat $k=2s$ and $c=3k/(2+k)$.

These predictions can be tested against exact results from the Bethe ansatz solutions of integrable spin Hamiltonians. An integrable Hamiltonian exists for each spin magnitude[9]; it is a degree 2s polynomial in the product of nearest neighbour spin vectors. Only for $s=1/2$ is this Hamiltonian realistic. However the critical theory discussed above is expected to be universal and to apply to a wide class of Hamiltonians of spin s. The Bethe ansatz gives the velocity parameter v, the slope of the specific heat at low T and the T=0 susceptibility. The parameters c and k apparently agree exactly with the above predictions.

The spin operators are represented in terms of fermions, in the continuum limit by

$$(2a)\mathbf{S}(x) \approx (\psi_L^+\sigma\psi_L + \psi_R^+\sigma\psi_R) + (-1)^{x/a}(\psi_L^+\sigma\psi_R + \psi_R^+\sigma\psi_L)$$

(sum over colour indices implied). The bosonized form after decoupling the charge and colour excitations is

$$2a\mathbf{S}(x) \approx (J_L + J_R) + \text{const}\cdot(-1)^{x/a}\text{tr}(g-g^+)\sigma$$

where g, an SU(2) matrix, is the fundamental field of the WZW model. Its anomalous dimension is $3/2(2+k)$. Thus the critical exponent determining the neutron scattering cross-section for momentum exchange near π/a is $\eta = 3/(2+2s)$.

These predictions can be tested by numerical diagonalization of finite chains. Conformal field theories defined on an line of length L with periodic boundary conditions, have energy levels forming conformal towers with spacing $2\pi/L$. These states can be labelled by their left and right SU(2) quantum numbers, $\mathbf{S}_L$ and $\mathbf{S}_R$ [These two SU(2)'s are only an exact symmetry of the critical theory. At finite length scales there is only a single SU(2) $\mathbf{S} = \mathbf{S}_L + \mathbf{S}_R$.] In the WZW model their is a primary operator (and hence a conformal tower of states) with $s_L=s_R=0,1/2,1,...k/2$. The lowest energy state in the tower has energy $E=(2\pi/L)2s_L(s_L+1)/(2+k)$. The degeneracies of states at each level in the towers for all values of s_L,s_R have been determined from the Kac-Moody algebra[5]. These energies should give the excitation energies (scaled by v) of states of quantum spin chains up to corrections which vanish as $L\to\infty$. Unfortunately, these corrections vanish very slowly. The reason is that a marginally irrelevant operator appears when deriving the WZW critical theory starting from the spin chain. This is $\lambda \mathbf{J}_L\cdot\mathbf{J}_R$. It was shown recently by Cardy[9] that marginally irrelevant operators lead to 1/LlnL corrections to the evergy gaps, and he expressed the coefficients of these correction terms in terms of β-function and operator product expansion coefficients. This behaviour arises because first order perturbation theory leads to a correction proportional to λ and renormalization effects replace λ the effective coupling at scale L $\lambda(L) \approx \lambda(L_0)/[1+\beta\lambda(L_0)\ln(L/L_0)]$. Thus the size of the correction for a small chain depends on the bare value of the coupling λ, but for large enough chains the correction has the universal form $1/\beta\ln L$. Applying Cardy's formula to the case at hand we obtain, for primary states with respect to the Virasoro algebra (these are the lowest energy states of given s_L,s_R in each tower)

$$\delta E = -\pi \mathbf{S}_L\cdot\mathbf{S}_R/L\ln L$$

For all k, the first excited multiplet has $s_L=s_R=1/2$. The above formula shows that the triplet is split below the singlet. This formula, and the $O(1/L)$ energy gaps obtained from the k=1 Kac-Moody algebra, agree exactly with recent Bethe ansatz calcuations for the s=1/2 chain[10].

The most interesting test of this theory is comparison with experiment. However, in order to make a comparison, it is neccessary to consider the effect of planar anisotropy. Intuitively, one might expect all massless degrees of freedom to become massive except for a single free boson associated with the unbroken U(1) phase. If this is the case, then c crosses over from $3s/(1+s)$ to 1. The remaining massless boson, φ, can be associated with the unbroken current

$$J^z = \sqrt{k/2\pi}\,\partial_-\varphi.$$

The other currents and the fundamental field can then be expressed in terms of φ and other commuting operators which are expected to become massive due to the anisotropy. In fact the same decomposition was discussed by Fateev and Zamolodchikov[11] who concluded that the commuting fields could be identified with the Z_k clock model. The conclusion is that the susceptibility varies continuously with anisotropy, since it is expressed completely in terms of φ. The exponent η jumps from $3/(2+2s)$ to $1/2s$. Thereafter the parameter k appearing in the susceptibility (for magnetic field directed along the z-axis) increases continuously and the exponent η (for the in-plane correlation function) decreases continuously obeying $\eta = 1/k$. c remains equal to one.

Experimental measurements of v (for an s=5/2 system), c (for s=1/2), k (for s=3/2 and 5/2) and η (for s=5/2 from quasi-elastic and inelastic neutron scattering) are shown in the figures. All predictions seem to be in reasonable agreement with the theory.

REFERENCES

1. I. Affleck, Princeton preprint, September, 1986.
2. E. Witten, Comm. Math. Phys. 92, 455 (1984).
3. V. Knizhnik and A. Zamolodchikov Nucl. Phys. B247, 83 (1984).
4. D. Gepner and E. Witten, Princeton preprint, 1986.
5. D. Gepner, private communication.
6. H.W. Blote, J.L. Cardy and M.P. Nightingale, Phys. Rev. Lett. 56, 742 (1986); I. Affleck, ibid. 56, 746 (1986).
7. J. L. Cardy, J. Phys. A17, L385 (1984).
8. A. Luther and I. Peschel, Phys. Rev. B9, 2911 (1974).
9. J.L. Cardy, UCSB preprint, 1986.
10. F. Woynarovich and P. Eckle, Freie Universitat, Berlin preprint, 1986.
11. V.A. Fateev and A.B. Zamolodchikov, Sov. Phys. JETP 62, 215 (1985).
12. P. Kulish and E. Sklyanin, Lecture Notes in Physics 151, 61 (1982); P. Kulish, N. Yu. Reshetikhin and E. Sklyanin, Lett. Math. Phys. 5, 393 (1981); L. Takhtajan, Phys. Lett. 90A, 479 (1982); J. Babudjian, Phys. Lett. 90A, 479 (1982); Nucl. Phys. B215, 317 (1983).
13. K.Takeda, S. Matsukawa and T. Haseda, J. Phys. Soc. Japan 30, 1330 (1971).
14. J.C. Bonner and M.E. Fisher, Phys. Rev.
15. G. Shirane and R.J. Birgeneau, Physica 86-88B, 639 (1977).
16. M. Niel, C. Cros, G. Le Flem, M. Pouchard and P. Hagenmuller Physica 86-88B, 702 (1977).
17. L.R. Walker, R.E. Dietz, K. Andres and S. Darack, Sol. State. Comm. 11, 593 (1972).
18. R.J. Birgeneau, R. Dingle, M.T. Hutchings, G. Shirane and S.L. Holt, Phys. Rev. Lett.26, 718, (1971).
19. M.T. Hutchings, G. Shirane, R.J. Birgeneau and S.L. Holt, Phys. Rev. B5, 1999 (1972).

FIGURE CAPTIONS

1. Specific heat of CPC. The dots are experimental data[13] and the solid curve is the result[14] of exact diagonalization of a finite chain. Linear behavior is seen at low T.
2. Dispersion relation[15] for TMMC showing sinusoidal behavior and $v/a = 70.7K$.
3. Susceptibility[16] of $CsVCl_3$ [the contribution from orbital angular momentum (about 91×10^{-6}uem CGS) has been subtracted]. The dot represents $k = 3$, the prediction for an isotropic $s = 3/2$ system.
4. Susceptibility[17] of TMMC. The point marked on the χ-axis corresponds to $k = 5$.
5. Quasi-elastic neutron scattering cross-section for TMMC. Q is a shifted, rescaled momentum: $Q \equiv (qa/\pi - 1)$. The points are experimental data[18] and the solid line is the theoretical prediction (with $\eta = .16 = 1/k$) smeared by a Gaussian to account for experimental resolution.
6. Peak neutron scattering cross-section[18] ($Q=0$) versus T. Points are experimental data and the line is the theoretical prediction $\propto T^{(1-\eta)}$ with $\eta = .16$.
7. Inelastic neutron scattering cross-section in TMMC. The points are experimental data[19] and the solid lines are theoretical predictions from with $\eta = .16$. Following the instructions in Ref. (19) the curves have been corrected for experimental resolution and acceptance. All four graphs have been fit with a single overall vertical scale factor so the theory predicts the q and T dependence of the peak heights.
8. Higher Q inelastic neutron scattering data in TMMC[15,19]. The theoretical curves have been corrected for experimental resolution following the instructions in Ref. (19). The normalization in not the same for the two different Q values so the scales were fit independently at each Q. The theoretical curves at $Q = .1$ were shifted to the left to account for the non-linearity of the dispersion curve.

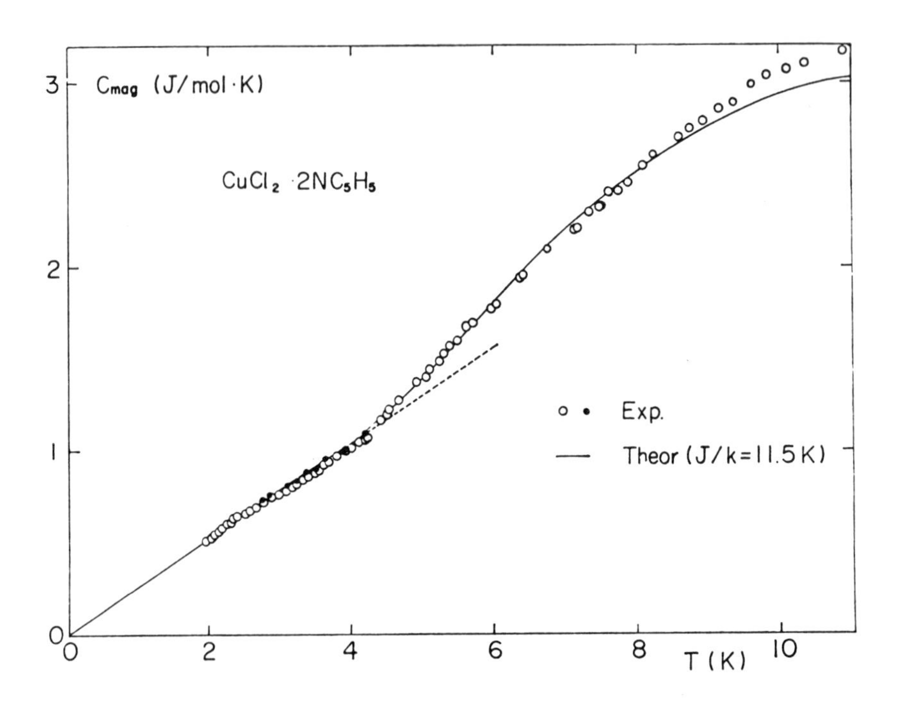

Fig. 1

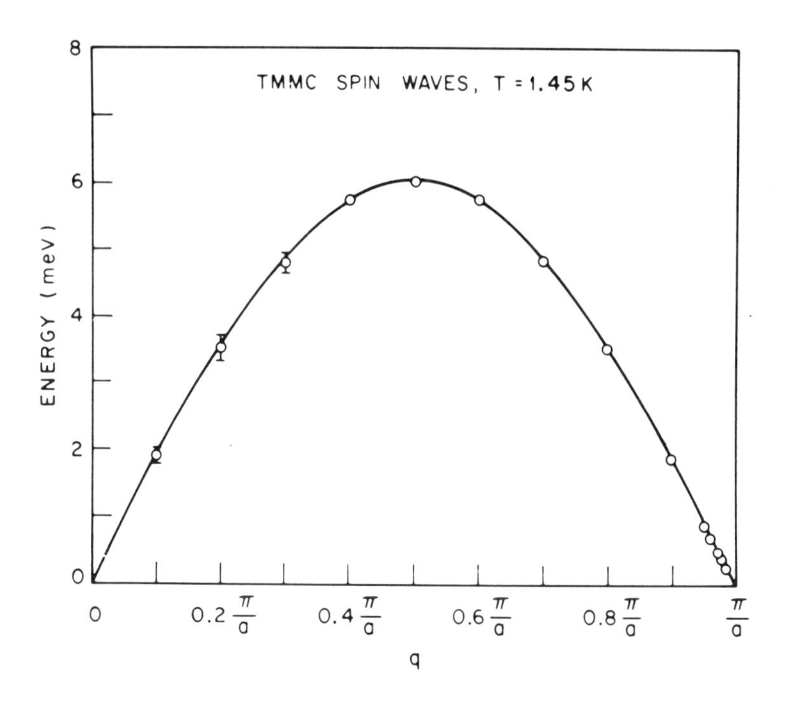
TMMC SPIN WAVES, T = 1.45 K
ENERGY (meV)
8
6
4
2
0
0
0.2 π/a
0.4 π/a
0.6 π/a
0.8 π/a
π/a
q

Fig. 2

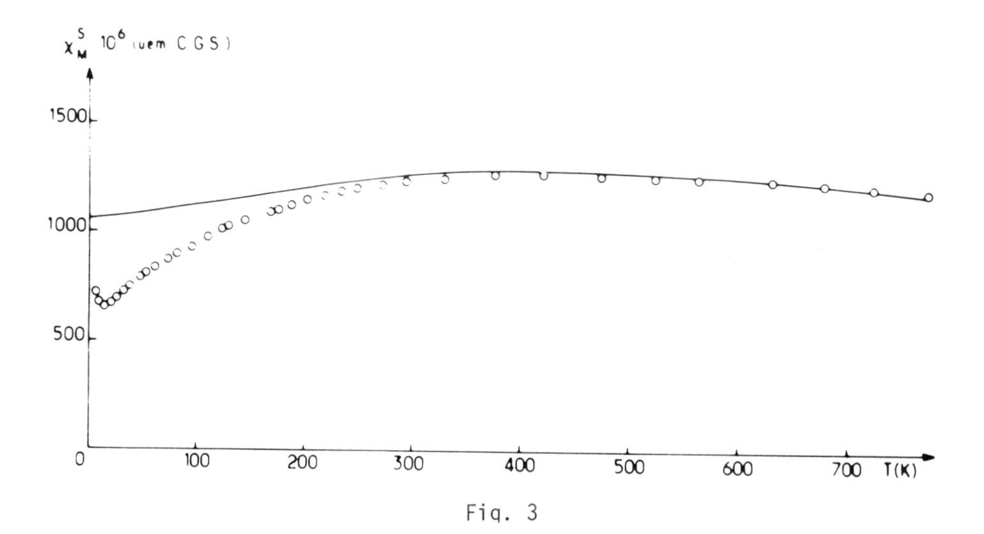
χ M S 10 6 (uem C G S)
1500
1000
500
0
100
200
300
400
500
600
700
T(K)

Fig. 3

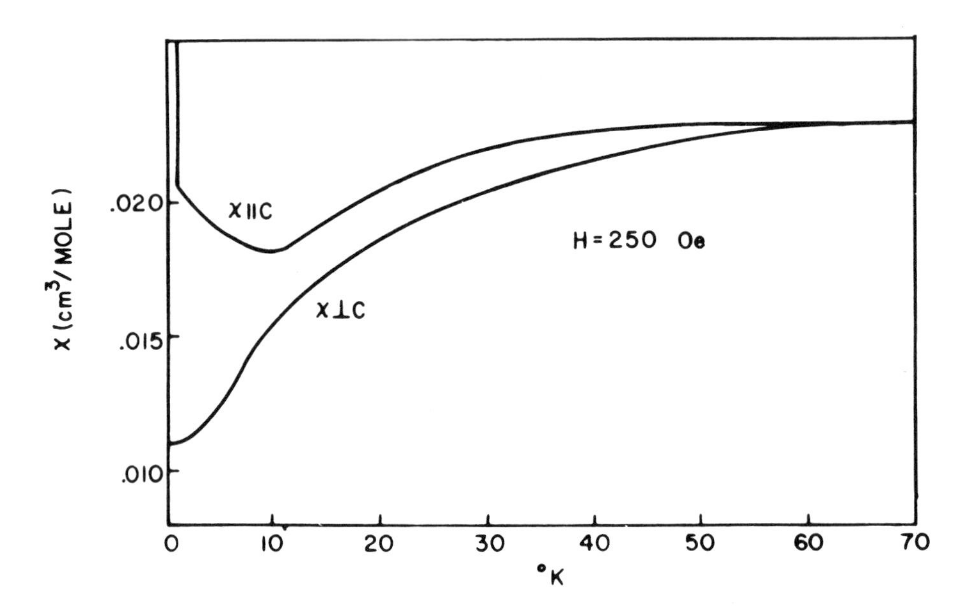

Fig. 4

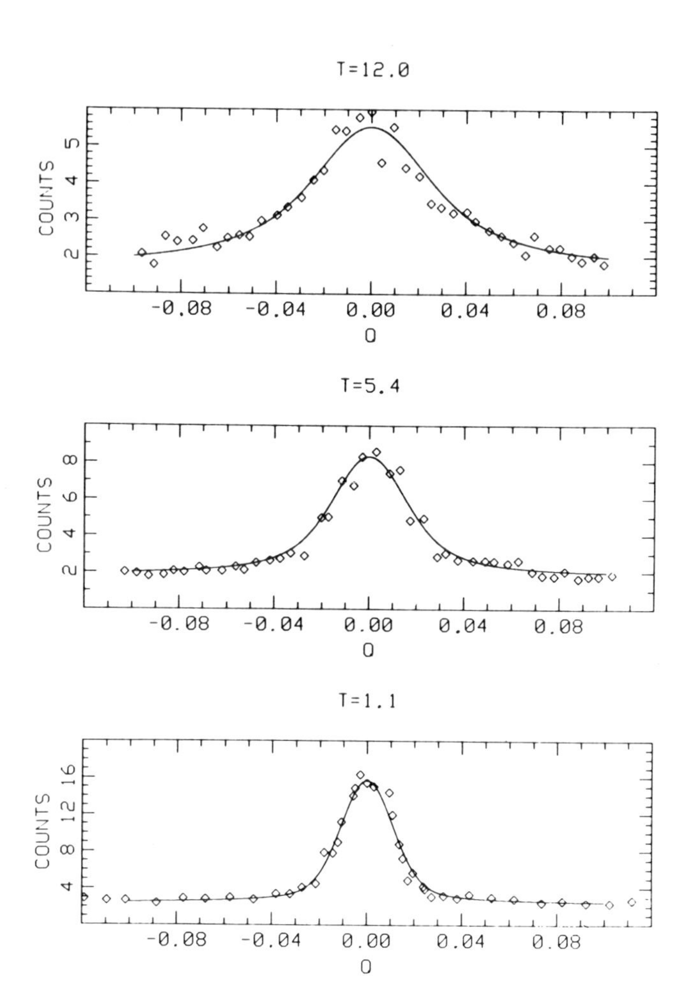
T=12.0
COUNTS
-0.08
-0.04
0.00
0.04
0.08
Q
T=5.4
COUNTS
-0.08
-0.04
0.00
0.04
0.08
Q
T=1.1
COUNTS
-0.08
-0.04
0.00
0.04
0.08
Q

Fig. 5

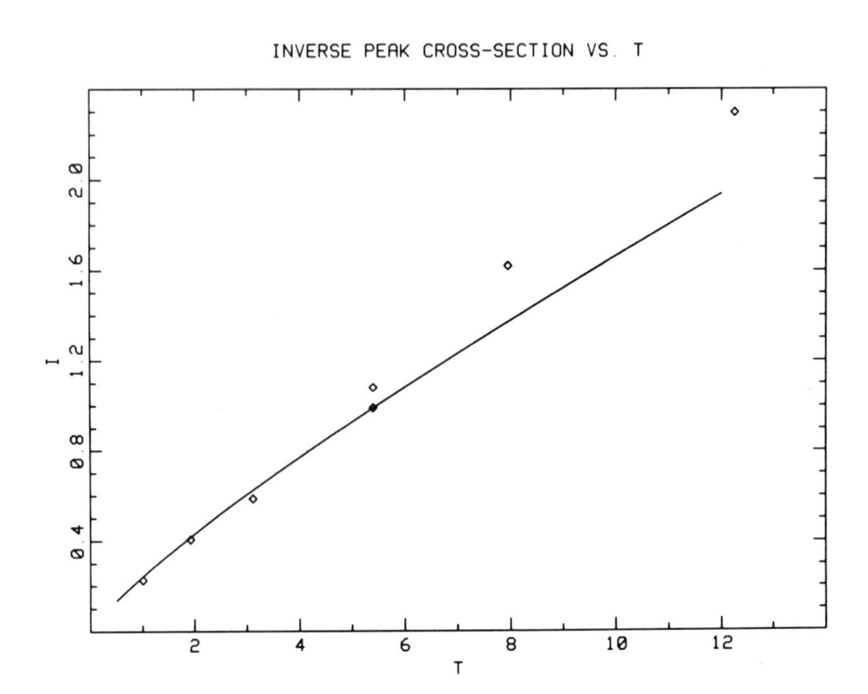

Fig. 6

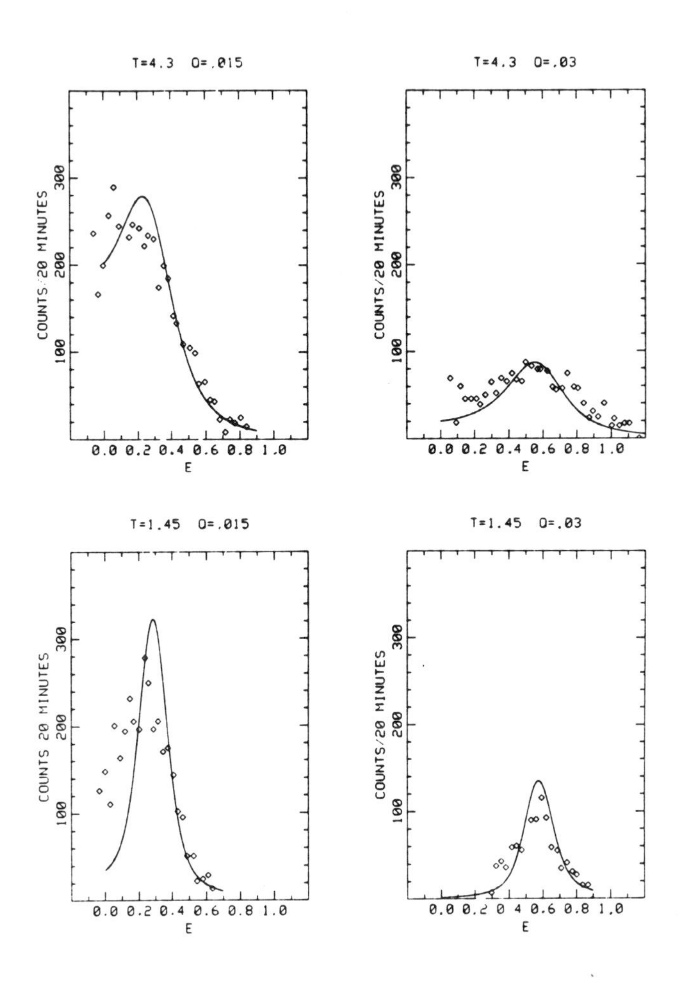
T=4.3 Q=.015
COUNTS/20 MINUTES
100
200
300
0.0 0.2 0.4 0.6 0.8 1.0
E
T=4.3 Q=.03
COUNTS/20 MINUTES
100
200
300
0.0 0.2 0.4 0.6 0.8 1.0
E
T=1.45 Q=.015
COUNTS 20 MINUTES
100
200
300
0.0 0.2 0.4 0.6 0.8 1.0
E
T=1.45 Q=.03
COUNTS/20 MINUTES
100
200
300
0.0 0.2 0.4 0.6 0.8 1.0
E

Fig. 7

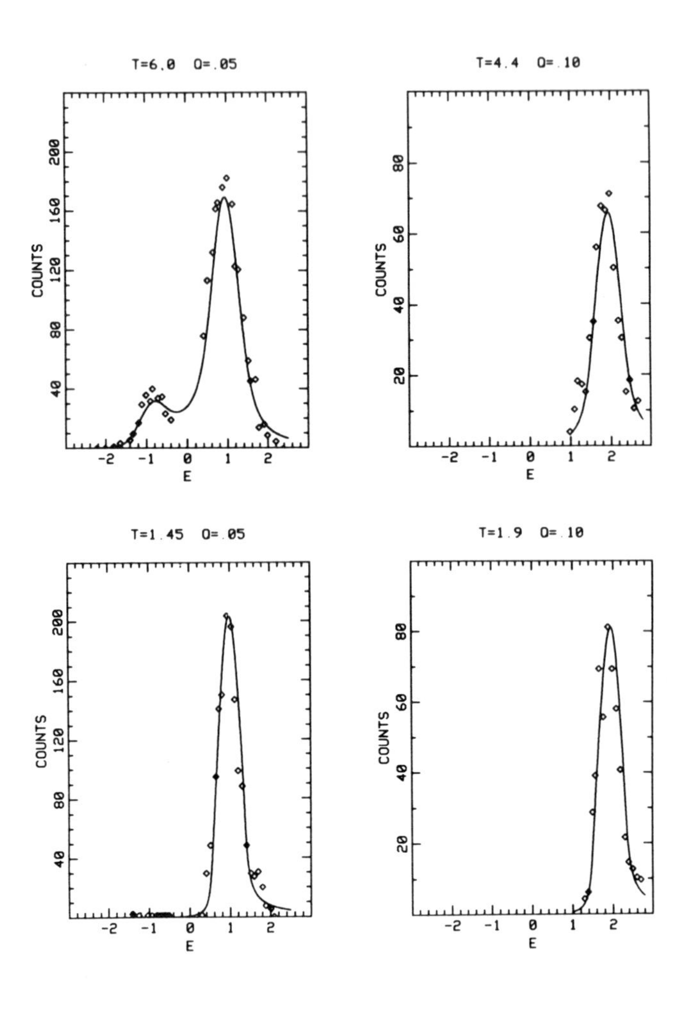
T=6.0 Q=.05
COUNTS
E
T=4.4 Q=.10
COUNTS
E
T=1.45 Q=.05
COUNTS
E
T=1.9 Q=.10
COUNTS
E

Fig. 8

CRITICAL PROPERTIES OF ADSORBED ATOMS AND CRYSTAL SURFACES: POSSIBLE REALIZATIONS OF MODELS OF INTEREST IN CONFORMAL INVARIANCE

Theodore L. Einstein
Department of Physics and Astronomy
University of Maryland
College Park, Maryland 20742 USA

1. INTRODUCTORY REMARKS

Adsorbed atoms on crystalline surfaces provide a rare example in nature of strictly two-dimensional systems. Moreover, due to the periodic potential of the substrate, such atoms tend to bind to a lattice of sites, making a lattice gas description often a reasonable approximation. In this presentation, we discuss how and to what extent such systems can be used to realize some of the predictions of conformal invariance.

In Section 2, we mention the important energies regarding adsorbed atoms and how they underlie the ordered patterns that are found. After a brief review of key terms, especially critical exponents, to describe 2-d lattice gases, we list the types of critical behavior to be expected and sketch the connection between lattice and magnetic models. In Section 3 we summarize our numerical work, mostly Monte Carlo, to suggest what one can reasonably expect to learn from experiments. This section introduces much of the formalism and concepts used to discuss the experiments in Section 4; the reader who is weary of the details should at least skim this section. Section 4 begins with a summary of the length scales involved. Specific heat and related measurements are presented first, followed by scattering experiments. The discussion is extensive, trying to show the scope of progress, but does not claim to be complete. In Section 5, we briefly cover some other experimental contacts with ideas of conformal invariance that are not presently feasible, although intriguing.

2. ADSORBED ATOMS AS LATTICE GASES

In describing adsorbed atoms (adatoms), it is convenient to speak of three characteristic energies: E_{AS}, the binding energy to the substrate; E_D, the diffusion barrier for adatom motion, i.e. the energy difference between being at the potential minimum (typically a high-symmetry site) and at a saddle point on a path between two such minima and E_{AA}, the interaction energy between two adatoms, i.e. the energy differences between two adatoms relatively close together and infinitely separated.[1] It is also useful (though not always entirely valid) to divide adsorption into two types: physisorption and chemisorption.

In physisorption, the adatom is bound to the surface by a physical, i.e. van der Waals, bond: $E_{AS} \sim 10^{-2}$eV. The interadsorbate interaction is also due primarily to van der Waals forces, so $E_{AA} \sim E_{AS}$ and is both isotropic and uniformly attractive for separations greater than the hard-core repulsion. Since $E_D \ll E_{AS}$, the adatom is relatively free to "skate" across the surface. Nonetheless, the periodicity of the substrate provides a field that breaks the two-dimensional translation invariance. The weak minimum, however, lead to large excursions which may introduce extra complications (e.g. changing real parameters to complex) which can change the universality class of the transition.[2],[3] Since $k_BT_c \sim E_{AA}$ and $E_{AA} \sim E_{AS}$, one expects notable desorption at temperatures T near the transition (at T_c). These experiments are thus performed in equilibrium with the gas phase.

In chemisorption the adatom is chemically bound to the substrate, with $E_{AS} \sim 10^0$eV. Such bonds can be either covalent or ionic in origin. The adsorption tends to be quite site-specific, with $E_D \lesssim E_{AS}$, although for adatoms with a single bonding orbital (esp. H) and close-packed surfaces (e.g. (111) faces of fcc crystals), $E_D \ll E_{AS}$ in certain directions, a complication we will ignore. For ionic bonding, the adatom interactions are direct, isotropic dipole-dipole repulsions and decay as r^{-3}, where r is the adatom-adatom separation. For covalent bonding, once the adatoms are separated by more than one or two sites, their interaction is predominantly "indirect," i.e. through the substrate via coupling to extended electronic states. When these states are anisotropic, as for transition-metal substrates, so is the interaction. It is also oscillatory in sign and decays rapidly with r: r^{-5} at large range, and often exponentially at closer range. In either case, $E_{AS} \gtrsim E_D > E_{AA}$. Near T_c there usually is little desorption; one has a strictly two-dimensional system with constant "coverage", θ.

For this presentation, the reader can simply imagine the interaction energies as initial givens, with the implicit understanding that they fall off rapidly with separation and need not be circularly symmetric. The ordered overlayer produced by these interactions are described in terms of the 2-d "primitive" vectors of the top layer of the substrate. For instance, on a square substrate with nearest-neighbor and next-nearest-neighbor (diagonal) repulsions, one finds at low enough T a (2x2) overlayer around a fractional occupancy or "coverage" $\theta = 1/4$. [With an attractive third-neighbor interaction, ordered islands will form at lower θ.] With just nearest neighbor repulsions, a checkerboard pattern of alternate occupancy occurs near $\theta = 1/2$. While this structure might be called $(\sqrt{2}x\sqrt{2})R45^{\circ}$, it is usually labeled c(2x2), i.e. centered (2x2) or a (2x2) with an extra adatom in the center. [The preceding (2x2) is often called p(2x2), p denoting primitive.] On triangular lattices of adsorption sites (e.g. on close-packed faces of fcc or hcp crystals), nearest-neighbor repulsions leads to a $(\sqrt{3}x\sqrt{3})R30^{\circ}$ around $\theta = 1/3$ (see Fig. 1 below). This pattern is widely found in physisorbed noble gas atoms on [the basal plane of] graphite. With a second-neighbor repulsion as well, a p(2x2) overlayer forms around $\theta = 1/4$.

These patterns are typically detected using low-energy electron diffraction (LEED),[4] although atom scattering,[5] low-energy positron diffraction,[6] etc. could also be used. In this energy range (20-300 eV) the mean free path before inelastic scattering is short, $\lesssim$ 10Å, making electrons very surface sensitive.[4] In this elastic scattering process, only two-dimensional crystal momentum is conserved: $\vec{k}_{\parallel}{}' = \vec{k}_{\parallel} + \vec{g}$, where $\vec{g}$ is a surface (or 2-d) reciprocal lattice vector (i.e. a member of the discrete Fourier series of the periodic lattice.) When an ordered overlayer is present, the real-space periodicity of the adatoms is larger than that of the substrate, so in reciprocal space new diffraction spots appear, as we shall explicitly see shortly. Because of the strong electron-atom interaction, the scattered intensity depends on multiple-site correlation functions. Most LEED analyses study intensity vs. incident electron energy spectra, with the goal of extracting the position of the adatoms relative to the substrate; data is fit basically by trial-and-error, using complicated computer codes to treat the multiple scattering. Instead we use LEED for what it is most naturally suited: to study the degree of order. We consider the temperature dependence of the scattered intensity and its profile in $\vec{k}$-space. For this problem, multiple scattering corrections are less important, and most of what follows will just consider single scattering (the "kinematic" approximation).[7]

In conformal invariance one speaks in terms of the anomalous scaling dimensions $\eta = [2-d] + 2x$ of various correlation functions at criticality. In general it is not possible to measure these directly[8]; we shall glimpse the difficulty using simulated data. Instead, one measures critical properties of various so-called densities as a function of their conjugate fields.[9] The densities of particular interest are the local order parameter, ψ, and the energy density ϵ. The order parameter has a dimensionality called n and is usually normalized to have unit amplitude at zero temperature T. For example, n = 1 for the Ising model (where ψ is the normalized coarse-grained sum of spins) or for binary fluids (where $\psi \propto \Delta\rho$, the difference in fluid densities). For a superfluid, n = 2: $|\psi|$ is the ratio of superfluid to total density, but there is an additional phase factor. For liquid crystals, ψ is a tensor. In magnetic systems, n is the dimension of the spins.

In surface experiments and most others, the field over which one has best (sometimes only) control is the reduced temperature $t \equiv |T-T_c|/T_c$, conjugate to the energy density. The long-range order or magnetization exponent β is defined by

$$|\langle\psi\rangle| \sim \frac{\partial f}{\partial h} \propto (-t)^{\beta}, \qquad T < T_c \tag{1}$$

Here f is the [reduced] free energy and h suggests the magnetic realization, with ψ the local magnetization and h the conjugate magnetic field. The susceptibility exponent γ describes how fluctuations of the order parameter (about its mean value) diverge near criticality:

$$\chi \sim \frac{\partial\langle\psi\rangle}{\partial h} \sim \frac{\partial^2 f}{\partial h^2} \sim \sum_{i,j} \langle(\psi_i-\langle\psi\rangle)(\psi_j-\langle\psi\rangle)\rangle \sim \chi_{\pm}|t|^{-\gamma^{\pm}}, \ \pm \equiv \text{sgn}\ (T_c\text{-}T\ ) \tag{2}$$

The correlation function $\Gamma_{ij} = r_{ij}^{-\eta_\sigma} f(r_{ij}/\xi)$, where ξ^{-1} corresponds to the mass gap in field theory and is called the correlation length and $f(o) = \text{const}$, $f(y) \underset{y\to\infty}{\sim} e^{-y}$. Near T_c, ξ diverges as

$$\xi = \xi_{\pm}\ |t|^{-\nu_{\pm}} \tag{3}$$

The specific heat exponent α describes the divergence of energy fluctuations.

$$C \sim \partial E/\partial T = C_{\pm}|t|^{-\alpha_{\pm}} + \text{const.} \tag{4}$$

Again one can express the energy fluctuation correlation function in a scaling form with an exponent η_ϵ.

According to the scaling hypothesis ξ is the only relevant length for determining the critical behavior, i.e. all the T dependence of the scaling function comes from ξ.[9] As consequence, $\alpha_+ = \alpha_- = \alpha$, $\gamma_+ = \gamma_- = \gamma$, $\nu_+ = \nu_- = \nu$. (The $\pm$ notation is unconventional: λ_- is denoted λ' while λ_+ is just λ). Moreover, from a fluctuation-dissipation equation, these exponents are related to the more familiar anomalous dimensional at T_c: $2-\eta_\sigma = \gamma/\nu$, $2-\eta_\epsilon = \alpha/\nu$, etc. In addition there are hyperscaling relations, involving dimension d and justified by invariance of the [reduced] free energy under scale changes:

$$d\nu = 2-\alpha = 2\beta + \gamma \tag{5}$$

Assembling these results, we find only two independent exponents in a universality class. From conformal invariance, just a single number, m or $c=1-6/[m(m+1)]$ characterizes a universality class, although more than one class may have the same c. We also note that the critical amplitude ratios C_+/C_-, ξ_+/ξ_-, and χ_+/χ_- are determined by the universality class. By duality[10], the first is unity for the $q = 2,3,4$ Potts systems of particular relevance to surfaces, to be described shortly.

Another important caveat is that these exponents are defined with the fields as the independent variables.[11] In adsorption experiments, this is often not possible for the equivalent of h. When a density, here θ, that would vary critically like ϵ at a transition is constrained, the exponents will change: β, γ, and ν are augmented by the factor $1/(1-\alpha)$, while $\alpha \rightarrow -\alpha/(1-\alpha)$. This "Fisher renormalization"[12] is expected when one crosses the phase boundary away from the saturation θ.

For tricritical points, and for multicritical phenomena in general, the situation is more complicated since there are more than two relevant fields. Recalling equations (1) and (2), there are different susceptibilities associated with the added fields, and so additional β's and γ's, as well as other exponents. These in turn can be readily related to the anomalous dimensions calculated in conformal invariance. Detailed expositions are available.[13],[14]

Before presenting the possible universality classes, it is worthwhile to review the common magnetic models since it is their names that grace the universality class of 2-d systems. We imagine a lattice of n-component spins on each site, with an inner product between the spins on nearest-neighbor sites:

$$\mathcal{H} = -J \sum_{\langle ij \rangle} S_i \cdot S_j \tag{6}$$

If these spins have continuous symmetry in spin space, these models are called O(n); more specifically O(1) is the famous Ising model, O(2) is the XY model, and O(3) is the Heisenberg model. There may also be anisotropy corrections of the form $w \sum_{\mu=1}^{n} (S_i^{\mu})^4$. For the

XY model, this fourfold anisotropy is called cubic. For the Heisenberg model, w >(<) 0 is called face- (corner-) centered anisotropy. Alternatively, the spins might be allowed to point in only q = n+1 discrete (symmetry) directions. The resultant q-state Potts model could also be written

$$\mathcal{H} = -J \sum_{<ij>} \delta_{S_i, S_j} , \qquad (7)$$

showing that these states could be called colors. The q=2 Potts model is just the Ising model. For q=2,3 but not larger, these are the Z_q models.

The connection between the overlayers and these magnetic systems is often not obvious initially. In general it is not possible to directly map the overlayer models into the magnetic ones: the lattice models have higher-order gradient terms. To relate the m-th neighbor E_m of the lattice gas to the exchange constant J, one must resort to a rather ambiguous mapping of configurations called "prefacing".[15]

Domany, Schick, and coworkers[16],[17] have used Landau arguments to propose that [only] four universality classes are possible for the critical melting transitions of adatoms, as indicated in Table 1. The possible realizations are listed in each column. At the top are the measurable exponents. For the XY model with cubic anisotropy, the exponents are non-universal. Notice that the exponent β varies very little between classes, so that it probably is better to use α or the exponents associated with critical scattering to distinguish classes. Details of the classification scheme, along with explicit forms of the Landau-Ginzburg-Wilson (LGW) Hamiltonian, can be found in Schick's review[17],[11].

In general, as just noted, the connections between the lattice gas systems and their magnetic analogues are subtle. An exception is the relation between a c(2x2) on a square net and an [antiferromagnetic] Ising model. If one writes $\mathcal{H}_{c(2x2)} = E_1 \sum_{<ij>} n_i n_j$ $(n_i=0,1)$ and $\mathcal{H}^{AF}_{Ising} = J \sum_{<ij>} \sigma_i \sigma_k$, $(\sigma_i=\pm 1)$, then the transformation

Table 1: Possible Continuous Transitions, adapted from Ref. 17.

Examples Discussed	Adsorption Site Symmetry \ UNIVERSALITY CLASS AND EXPONENTS	ISING	x-y WITH CUBIC ANISOTROPY	3-STATE POTTS	4-STATE POTTS
	α	0 (log)	NON UNIVERSAL	1/3	2/3
	β	1/8		1/9	1/12
	γ	7/4		13/9	7/6
	ν	1		5/6	2/3
Ni(110) Au(110) W(112)	SKEW (p1) OR RECTANGULAR (p2mm)	(2×1) (1×2) c(2×2)			
Ni(100) W(100) W(110)	CENTERED RECTANGULAR (c2mm) OR SQUARE (p4mm)	c(2×2)	(2×2) (1×2) (2×1)		
graphite	TRIANGULAR (p6mm)			$(\sqrt{3}\times\sqrt{3})$	(2×2)
	HONEYCOMB (p6mm)	(1×1)			(2×2)
Ni(111)	HONEYCOMB IN A CRYSTAL FIELD (p3m1)			$(\sqrt{3}\times\sqrt{3})$	(2×2)

$\pm\sigma_i \Leftrightarrow 2n_i - 1$ takes one model to the other if we demand that $\Sigma\, n_\ell = N/2$ in the lattice gas, i.e. that half the sites be occupied. The $\pm$ refer to the two "colors" of sites; if we imagine the lattice as a checkerboard, $\pm \rightarrow (-1)^{\ell_x+\ell_y}$. In this special case $E_1 = 4J$. Moreover, we see that the staggered magnetization, the order parameter becomes

$$\psi = (2/N)\, \Sigma\, (-1)^{\ell_x+\ell_y} n_{\vec{\ell}} = \frac{2}{N}\, \Sigma\, e^{i2\pi(\frac{1}{2},\frac{1}{2})\cdot(\ell_x,\ell_y)} n_{\vec{\ell}}. \tag{8}$$

Thus the $\vec{k}$ point in reciprocal space associated with the ordering corresponds to the corner of the Brillouin zone (BZ), i.e. the proximity cell of the lattice of reciprocal points $\vec{g}$ in $\vec{k}$ space. (The reciprocal lattice and BZ have the same point group symmetries as the corresponding real-space lattice.) Landau theory demands that such $\vec{k}$ points, called here $\vec{k}_o$, lie at high-symmetry points of the edge of the BZ.[15] (We have, however, studied a counterexample, which melts continuously to an incommensurate disordered phase.[18])

In Fig. 1 we suggest the connection between the $(\sqrt{3}\times\sqrt{3})R30^o$ lattice case and the 3-state Potts model.[19] The Potts symmetry in the lattice gas arises from the interaction between ordered domains

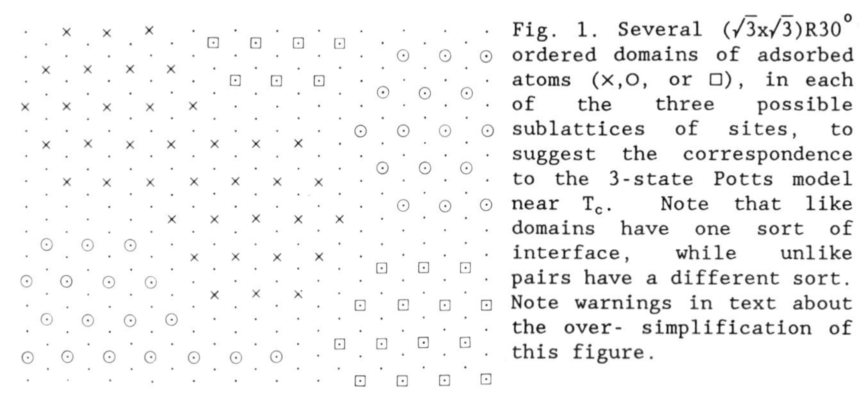

Fig. 1. Several $(\sqrt{3}x\sqrt{3})R30^{\circ}$ ordered domains of adsorbed atoms (×,○, or □), in each of the three possible sublattices of sites, to suggest the correspondence to the 3-state Potts model near T_c. Note that like domains have one sort of interface, while unlike pairs have a different sort. Note warnings in text about the over- simplification of this figure.

in the three degenerate ground states rather than from the microscopic interactions between the adsorbates. (The maximum size of these domains is of order ξ, but in contrast to the figure there will also be a hierarchy of smaller domains.) Like domains have one sort of interface (and so interaction), unlike (no matter which pairing) seem to have one different sort. (We shall see in section 3 that the latter statement is not quite true.)

3. RESULTS OF NUMERICAL SIMULATIONS

For diffraction experiments, most notably low-energy electron diffraction, it is useful to compute[20] the structure factor, i.e. the Fourier transform of the occupation-number pair correlation function

$$S(\vec{k},T) = \sum_{r,r'} e^{ik\cdot(r-r')} < n(r)n(r') > \equiv \langle | \sum_{r} e^{ik\cdot r} n(r)|^2 \rangle. \qquad (9)$$

In the limit of single scattering (called the kinematic approximation), the intensity of electrons scattered by $\vec{k}$ is proportional to $S(\vec{k},T)$. More generally, multiple scattering can significantly modify the amplitude of the critical properties we seek, but not the exponents. (cf. Section 5.)

If the longer-range order were associated with $\vec{k} = 0$, we could write

$$\begin{aligned} S(\vec{k},T) &= \mathcal{F}\{< n(r)n(r')>\} \\ &= \mathcal{F}\{<(n(r)-<n>)(n(r')-<n>)> + <n>^2\} \qquad (10) \\ &= \chi(k) + m^2\delta(k) \end{aligned}$$

where χ is the susceptibility and the "magnetization" $\langle n \rangle$ the long-range order. In fact, as suggested in the discussion of the c(2x2) overlayer, we are here concerned with $\psi(r) = n(r)e^{ik_o \cdot r}$, where k_o is the position in reciprocal space associated with long-range order. Then one can expand the scaling function (to be defined in equation (13)) to show, for small $k\xi$,

$$S(\vec{k},T) \sim \sum_{k_o,g} \frac{\chi(k_o+g,T)}{(k_o+g-k)^2+\xi^{-2}} + A\delta_{k,k_o+g}(-t)^{2\beta}\theta(-t) \tag{11}$$

For a particular ordered overlayer there in general are several inequivalent (that is, not connected by a substrate reciprocal lattice vector g) k_o's that go into each other under the point group operations of the lattice of binding sites. The asymptotic dependence of the long-range order part is written explicitly. The critical scattering part will be Lorentzian with a width $\propto |t|^{\nu}$ and a height $\propto |t|^{-\gamma}$. This diffraction picture is appropriate to the limit of a "perfect instrument," i.e. $\xi < L_i$, where L_i suggests the length over which the experiment samples the surface coherently. For typical diffraction instruments this length is of order 100Å, so that one might see a distance of say 40 sites.

A different expansion, derived first by Fisher and Langer,[21] applies in the limit $\xi > L_i$. Applied to scattering,[22] we find that in this limit the <u>integrated</u> intensity, integrated out to k_I from the center of an overlayer-induced spot, has the form

$$I(T) = A_o - A_1 t \mp B_{\pm}|t|^{1-\alpha_{\pm}} + \ldots, \tag{12}$$

where the first two terms come from the analytic contribution. For the Ising model, the coefficients can be evaluated analytically.[22] The origin of the 1-α anomoly is its presence in the correlation functions [at least in combinations from which one cannot deduce the phase of the order parameter], including multisite functions. It is remarkable that by doing a simpler low-resolution experiment, for which one does not need to take into account instrument response, one can measure the exponent α, which distinguishes most clearly between universality classes, certainly better than β.

According to the phenomenological theory of second order phase transitions,[9] the structure factor is expected to scale as

$$S(\vec{k},T) = a_1 t^{-\gamma} X_{\pm}(a_2 t^{-\nu} |\vec{k}|) \tag{13}$$

for small k and t. In the scaling function X(y) it is assumed that $\vec{k}$ is measured from k_o rather than the origin. The $X_{\pm}(y)$ are universal functions, while the a_i are system-dependent constants. Since $S(k \neq k_o, T_c)$ remains finite,

$$X_{\pm}(y) \sim y^{-\gamma/\nu} = y^{\eta-2} \text{ as } y \to \infty. \tag{14}$$

In order to see how far away from T_c (and k_o) these expressions are likely to hold for 2-d lattice gas systems, and to assess some of the complications that might arise, we performed explicit calculations for systems with $(\sqrt{3}x\sqrt{3})R30^o$ and p(2x2) order on triangular substrates having slightly below 4000 sites, comparable to the size of defect-free regions on metal surfaces. The boundaries of the lattices were hexagonally shaped to insure that as many of the infinite-system symmetries as possible were included: periodic boundary conditions were used to minimize finite-size and other system-dependent effects, even though they are clearly unrealistic for surfaces. The runs were relatively long (typically $2x10^5$, up to 10^6, Monte Carlo steps per site). Further details are in Ref. 20.

To check the viability of equation 14, we plotted log S(k) vs. log k about 3% above T_c. While the plots did appear linear, the effective η for the $\sqrt{3}$ case varies between 0.03 and 0.17, depending on how one draws a line through the points, in comparison with the expected value of $4/15 \approx 0.266$.[24] Similar values for the p(2x2) case, between 0.08 and 0.20 should be compared with $\eta = 1/4$ for the 4-state Potts value. For the Ising model above T_c, $S \propto y^{\eta-2}$ (η = 1/4) to an accuracy of 5% only when $y > 163$[8], while our data scales only up to $y \sim 10$, so it is unlikely that this approach is experimentally useful.

At T_c, where $\xi \sim L$, the system size, S satisfies another scaling relation:

$$S(k,T_c,L) = L^{\gamma/\nu} W(\vec{k}L). \tag{15}$$

Using this expression (noting $\lim_{y\to\infty} W(\vec{y}) \propto |\vec{y}|^{\eta-2}$) at our best estimates of T_c for the $\sqrt{3}x\sqrt{3}$ and p(2x2) lattices, we obtained η_{eff}= 0.23 and 0.30, respectively. Since the scaling function W depends on the geometry and boundary conditions, it is not obvious what to expect for realistic non-periodic boundaries and irregular shapes. For the Ising model on a square lattice, we[24] have numerically checked the predictions of conformal invariance for $S(k,T_c)$, derived by Kleban and coworkers,[25] and studied at what wave vectors the discrete-lattice effects appear. Monte Carlo was used to study square, circular, and 2x1 rectangular geometries with free boundary conditions which transfer matrix methods were used for a strip with periodic boundary conditions (i.e. or infinite cylinder). In all cases the data scaled well for $k \le \pi/2a$, and the amplitudes agreed nicely with the values deduced from conformal invariance.

In Fig. 2 the data is displayed plotted according to equation (13). The wave vector runs from the corner of the hexagonal BZ toward the center, with scaling holding, to within statistical error, half way. For larger $\vec{k}$, the lattice constant becomes important. Data within ~ 2% of T_c does not scale due to finite-size effects.

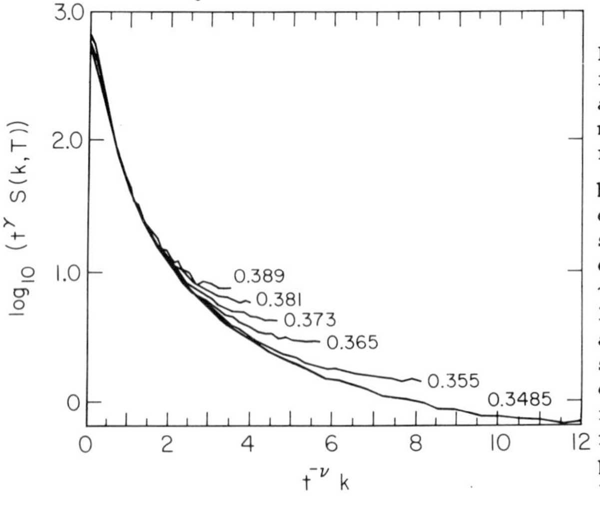

Fig. 2. Structure factor above T_c for a lattice gas with nearest-neighbor repulsion E_1=1, having a $(\sqrt{3}x\sqrt{3})R30^o$ ordered state, scaled according to eqn. (13), assuming γ and ν of the 3-state Potts model and T_c=0.338; $\vec{k}$ starts at the BZ corner, equidistant from three reciprocal lattice points, and goes toward one of them.

Quantities such as χ or ξ (or $|\psi|)^{-1}$ are expected to have corrections to scaling of the form

$$G(T) = G_o^{\pm}\, t^{-\lambda}(1+a_G^{\pm} t^{\Delta} + \ldots), \tag{16}$$

where the leading Δ is expected to be no greater than unity. (The arbitrary definition of t suggests the general presence of analytic corrections). Hence in an experiment, physical or Monte Carlo, one expects[20] to observe an <u>effective</u> exponent

$$\lambda_{eff} = -\frac{\partial \ln G}{\partial t^*} \sim \lambda - a_G^{\pm}\, \bar{t}^{\Delta}\Delta \tag{17}$$

averaged over some thermal range, and where t^* is an estimate of t. [Since T_c is non-universal, dependent on specifics of interactions, it is rarely known <u>a priori</u>.] While one can in principle explore the convergence properties of λ_{eff},[20] in typical experiments involving one or two decades of data, one can do little more than quote effective exponents. Even the choice of data range to fit involves some subtle but clearly important choices.[20]

We fit S(k,T) to a lorentzian to obtain ξ and χ (cf. equation 11), below T_c the point at $\vec{k} = 0$ must be excluded. Above T_c, we take $t \equiv 1 - T_c/T$. Fig. 3a shows the result of a fit to $\chi \propto t^{-\gamma}$ for the $\sqrt{3}x\sqrt{3}$ case. The [Monte Carlo] data are linear over about 1 1/2 decades, from 0.015 to 0.25; $\gamma_{eff} = 1.25 \pm 0.07$, 13% below the pure value 13/9 = 1.44. The log-log plot of ξ in Fig. 3b is linear over a similar range; $\gamma_{eff} = 0.77 \pm 0.05$, 7% below the pure value of $5/6 \approx 0.83$. [The error bars in this section, and generally in the next, are due to statistic and do not include possible systematic effects.]

Below T_c, the situation is more complicated. In Fig. 3c, a log-log plot of order parameter (squared) vs. t, with T_c fixed at the value deduced above the transition, shows clear non-linearities. No choice of T_c makes this plot linear over the whole fitting range. Evidently corrections to scaling are larger below than above T_c. Since the ratio of their amplitudes is expected to be universal, this problem should be general. If experiments could get closer to T_c,

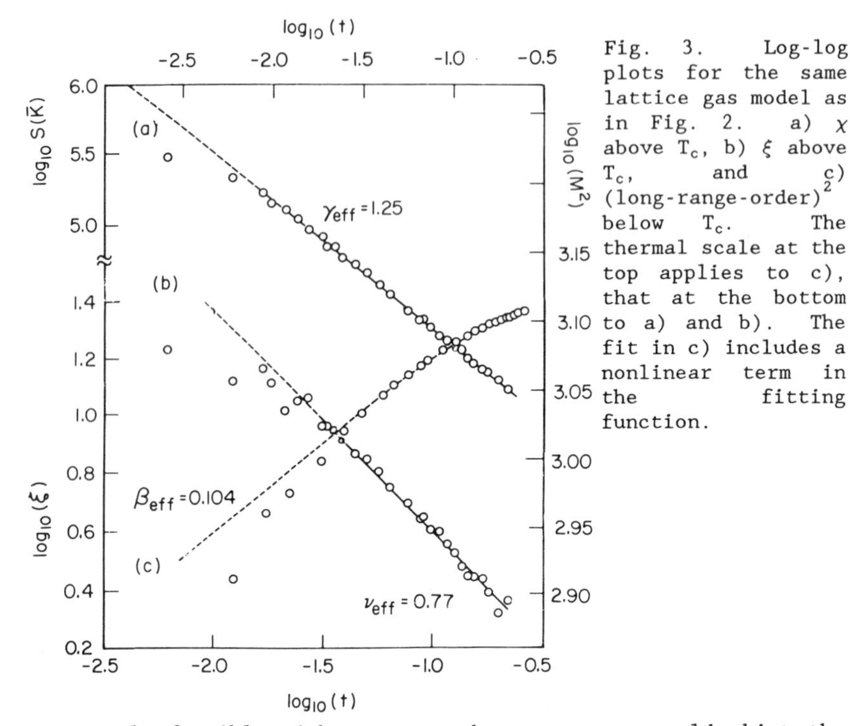

Fig. 3. Log-log plots for the same lattice gas model as in Fig. 2. a) χ above T_c, b) ξ above T_c, and c) $(\text{long-range-order})^2$ below T_c. The thermal scale at the top applies to c), that at the bottom to a) and b). The fit in c) includes a nonlinear term in the fitting function.

not remotely feasible with present substrates, one could skirt the issue by dropping data relatively far from T_c. While there is not enough data range to allow Δ as a separate fitting parameter, sometimes one has an idea of its value. As alluded earlier, one at least expects a non linear term in the thermal scaling field. Fitting the squared order parameter to $(t + bt^2)^{2\beta}$ as shown in Fig. 3c, adequately accounts for the data and gives $\beta_{eff} = 0.104$, 6% below the pure value of $1/9 \approx 0.111$. Here T_c was fixed at the value determined by fits above T_c. If it is also a free parameter, the best (minimum chi-squared) fit lowers T_c^* by 1%, while β plummets to 0.087. If T_c is chosen to maximize the linearity of the log-log plot, assuming no corrections to scaling, T_c drops slightly more and β_{eff} falls to 0.073. The reader is thus well advised to be skeptical of surface experiments considering only the long-range order as a function of temperature. For the 3-state Potts model, Nienhuis[26] predicted the smallest Δ is 2/3 (corresponding[27] to $x = 4/5 + 2$, using $2(x-1)\nu = 2 - \alpha + 2\Delta$). Using this value gives $\beta_{eff} = 0.111 \pm 0.019$, fortuitously exact agreement.

As $T \to T_c$, the ratio of ξ and of χ, above and below T_c, is universal. For the Ising model, the ratios $\xi_+/\xi_- = 3.16...$ (using the second moment definition) and $\chi_+/\chi_- \approx 38$.[28] For the $\sqrt{3}x\sqrt{3}$ lattice gas, these ratios are 3.2 ± 0.6 and 43 ± 12, respectively, with even larger systematic errors possible due to the notable corrections to scaling beneath T_c. Similar calculations for a 3-state Potts model on a lattice with comparable degrees of freedom gave ratios of 4.1 ± 0.2 and 43 ± 3 for $0.015 < t < 0.1$. The large ratio for χ shows that it will be extremely difficult to measure critical scattering beneath T_c. In modest-resolution diffraction experiments, there will be the further need to account for a convolution with an instrument response function when trying to separate critical scattering from Bragg-like scattering. Thus while we have computed γ' and ν',[20] we do not discuss them here; their physics is largely consistent with the previous discussion of β.

So far we have been comparing the $\sqrt{3}x\sqrt{3}$ lattice gas to the 3-state Potts model. A difference arises from the coupling of the order parameter degrees of freedom to the spatial ones of the lattice gas. In the $\sqrt{3}x\sqrt{3}$ case, since $\vec{k}_0$ occurs at the corners of the hexagonal-BZ, one only expects 3-fold rotational symmetry of $S(\vec{k},T)$ (rather than the 6-fold symmetry of the lattice). This behavior is observable in contour plots of $S(k,T)$. It corresponds to the following small-$\vec{k}$ expansion:

$$S(k,T) = \chi(T)[1-\xi^2(T)|\vec{k}|^2 + b(T)\ (k_r{}^3-3k_a{}^2k_r)\ ...], \tag{18}$$

where r and a denote the radial and azimuthal components. This asymmetry is due to a triaxial chiral term in the LGW Hamiltonian.[29] Physically, the domain-wall energy between, say, domain B to the left of domain A is different from that with A to the left of B. (The term chiral comes from the difference in energy between clockwise and counterclockwise ABC (3-domain) vertices. A closer look at Fig. 1 may clarify this, or see Ref. 29). If we assume that the anisotropy is a correction to scaling, then

$$S(k,T) = t^{-\gamma}[X_{\pm}(|\vec{k}|t^{-\nu}) + gt^{\Delta}Y_{\pm}(\vec{k}t^{-\nu})], \tag{19}$$

with $Y_{\pm}$ (universal) correction-to-scaling functions. Fitting S(k,T) to the form of equation 18, we obtain b(T). A log-log plot of b/ξ^3 vs t is linear; interpreted in terms of equation (19) yields $\Delta_{eff} = 0.83 \pm 0.20$. This result is consistent with den Nijs's prediction[30] that the (irrelevant) triaxial field leads to $\Delta = 5/6$ (corresponding[27] to x=3). Other interpretations of this behavior are also possible.[20]

The other lattice gas we considered, the p(2x2), has nearest and second neighbor sites vacant in the ordered states. The existence of four degenerate ordered states leads to a correspondence with the 4-state Potts model. Since the exponents are not quantized for this c=1 case, there is perhaps less interest in scrutinizing experimental examples as illustrations of conformal invariance. Above T_c, we find $\gamma_{eff} = 1.13 \pm 0.06$ and $\gamma_{eff} = 0.70 \pm 0.09$, a few percent below the 4-state Potts values of $7/6 \approx 1.17$ and $2/3 \approx 0.67$, respectively. Below T_c, nonlinearities are again inescapable in log-log plots. Including a non-linear term in the fit gives $\beta_{eff} = 0.083 \pm 0.009$, in agreement with the 4-state Potts value of 1/12. Since the q-state Potts model becomes first order for $q > 4$,[23] there is a marginal field (the dilution field) at $q = 4$, leading to logarithmic corrections to scaling. Using this information in the fit but freeing T_c allows T_c to drop slightly, but β_{eff} is 0.078 ± 0.020, very close to the value of 0.077 obtained with just a non-linear term. Here the additional a priori information does not help. Neglecting corrections to scaling completely leads to $\beta_{eff} = 0.066$, similar to previous Monte Carlo work.[31] In this lattice gas, $\vec{k}_o$ lies at the middle of a BZ boundary (half way between to $\vec{g}$ points), so only two-fold symmetry is expected. The new anisotropy term in the expansion of $S(\vec{k},T)$ is quadratic, and analysis of the oval-shaped contours is rather inconclusive beyond showing the greater potency of the gradient-squared addition to the LGW Hamiltonian.

For applications to conformal invariance, it would be worthwhile to perform similar simulations of examples of Ising and 3-state Potts tricritical points. We plan to carry out the former in conjunction with previously published transfer-matrix studies,[32] the germane results of which we briefly summarize: The density discontinuity at a first order phase transition can be computed with transfer matrices, using a generalization of Hamer's method,[33] involving matrix elements of the density operator. The tricritical point can then be located as the point where this jump [first] vanishes. As a test, we considered the exactly-solved hard-square model with second-neighbor attractions.[34] Extrapolating strip widths L = 12, 14, and 16 locates the tricritical point to within better than 0.02%. Applying the finite-size, periodic-strip scaling result[35]

$$\frac{\ln\lambda_1}{L} = f - \frac{\pi c}{6L^2} \tag{20}$$

(with λ_1 the leading eigenvalue) gives c = 0.70003, in excellent agreement with the exact result of 7/10.[27] Moreover, using the result that each anomalous scaling dimension can be related to a correlation length or mass gap $[(\eta_i = L/\pi\xi_i(T_t, L)]$, we obtained the five tabulated values[27] plus the first daughters in the conformal towers of the leading two.

4. EXPERIMENTAL ILLUSTRATIONS

4.1 Length Scales

Implicit in our Monte Carlo work is the limited size of defect-free flat regions on surfaces. Here we simple speak in terms of some characteristic length L_s, which limits the size of ξ and so the closeness of approach to T_c. The form of the distribution of sizes around this L_s is an unanswered question, with calculational fits involving very simplified models.[36] The boundary conditions at the defects, which may be steps (i.e. flat terraces meeting, with a lattice constant vertical separation) or isolated impurities or missing substrate atoms, are also ill-defined,[37] but certainly not like the ideal conditions in calculations, i.e. not periodic or free. If an adatom is too strongly bound at the impurity, then it renormalizes the impurity interaction. For most transition metals L_s ~ 150Å, so that the defect-free regions have 10^3-10^4 sites. These are the substrates used for chemisorption studies since they are hard and are "active" enough chemically to dissociate gas molecules (i.e. E_A is large). Semiconductor surfaces may also be chemically active, but the strongly covalent nature of their binding makes the cleaned surface generally unstable. The surface is likely to "reconstruct",[38] i.e. spontaneously form lower (2-d) symmetry structures to saturate the "dangling bonds" extending out from the freshly formed surface. Thus, in physisorption experiments one seeks a chemically inert substrate. Graphite is the common choice. This laminar material has most of its binding within the layers. In principle, it should thus be less prone to defect formation. Exfoliated graphite ("grafoil"), however, has L_s comparable to metals. The big advantage is that there is an enormous "surface"; the gas atoms need not just adsorb on the outside. This feature is very important for weakly scattering probes such as neutrons and x-rays or for most specific heat measurements, which would be overwhelmed by a metallic substrate. Improved forms of graphite have larger L_s: for ZYX, $L_s \sim 10^3$Å (with estimates of $\gtrsim$ 600Å[39],1400Å[40], 2000Å[41] and $\geq$ 3000Å[42] appearing); for graphite foam $L_s \sim 10^3$Å.[43] For HOPG (highly oriented pyrolytic graphite), used in many recent studies, $L_s \sim 10^4$Å.[36] Another problem with [exfoliated] graphite is that the microcrystals are not perfectly aligned, azimuthal and more importantly with respect to the direction of the normal axis; these problems are most severe for grafoil but are still considerable for the better graphite surfaces. Accordingly, more complicated, though apparently well-characterized fitting functions[41,42] than equation (11) are needed.

In diffraction experiments, one must also be concerned about the instrument resolution, discussed in section 3 in terms of a characteristic length L_i over which the sample is coherently probed. For conventional LEED or neutron scattering (off powder samples), $L_i \sim$ 100Å. High-resolution LEED systems increase this by a factor of 5. In some apparatuses, another factor of 2-4 can be achieved, but at a high cost in experimental difficulty.[44] A novel set-up involving a mirror electron microscope (MEMLEED)[45] promises L_i approaching 10^4Å, but has not yet come close to achieving this target.[46] Atom scattering has L_i roughly comparable to LEED; e.g. He scattering can under optimal conditions attain $L_i \sim 500Å$.[47] In general, atoms are more difficult to use than electrons, but have the feature of extreme surface sensitivity since they do not penetrate the substrate. At the other extreme are x-rays. While x-rays from a rotating anode (the highest - intensity lab source) have L_i comparable to conventional LEED, nearly two orders of magnitude improvement can be attained using a synchrotron. (But then the surface scientist must emulate the high-energy physicist, traveling to remote facilities and running intensively for short spells.)

4.2 Specific Heat and Related Measurements

Almost all the experiments to be discussed were reported during the 80's. In 1980, Tejwani et al.[48] measured the specific heat of ^{4}He on grafoil. Below $T_c \sim 2.9$K, a $(\sqrt{3}x\sqrt{3})R30^o$ ordered state exists. Log-log plots both above and below T_c of the specific heat [minus a background term] are linear for $\sim 5 \times 10^{-3} < t < 10^{-1}$, with $\alpha \approx 0.28$, consistent with $\alpha = 1/3$ for the 3-state Potts model. (If the data are massaged, even better agreement is possible.[49]) The He was removed and replaced by Kr. The Kr atoms also form a $(\sqrt{3}x\sqrt{3})R30^o$ with a much higher T_c. When He was readmitted into the chamber, one of the three equivalent ground state had been eliminated by this essentially quenched impurity. Since there were only two equivalent states to choose between, the symmetry changed to that of the Ising model. Indeed, α dropped dramatically, consistent with $\alpha \sim 0$ (logarithmic divergence), characterizing Ising behavior.

Chan and coworkers studied N_2 on graphite foam. For $\theta \gtrsim 1/3$ a $(\sqrt{3}x\sqrt{3})R30^o$ forms below ~86K via a continuous transition. For lower coverages (and temperatures), except close to saturation ~1/3 there is a coexistence regime between this ordered phase and a dilute disordered (so-called fluid) phase, indicative of a first-order line in the corresponding chemical potential-T plane. This behavior is suggestive of a strong second-neighbor attraction (as well as a nearest neighbor repulsion). Measurements of the vapor-pressure isotherms[43] and heat capacity were made.[50] At the intersection of the first order and continuous lines is a <u>tricritical</u> point, presumably in the 3-state Potts class, which occurs at $\theta_t \sim 0.34$ and

T_t = 85.37K. For $\theta \lesssim \theta_t$ there is a narrow sliver of coexistence regime, since the ordered phase persists (with rapidly falling transition temperature) down to $\theta = 1/3$. To analyze the data, the point of steepest increase of θ vs. pressure was located and then the pair of coverages at ± 1% (or ± 1/2%) of this vapor pressure were determined. Using either criterion, plots of reduced coverage vs. reduced pressure were found to scale. [For ± 2% or ± 3%, similar plots did not scale; this breakdown was attributed to heterogeneity in the binding energy E_A.] Log-log plots of the difference between each pair of coverages vs. $t \equiv (T_t-T)/T_t$ were found to be linear for $0.007 \lesssim t \lesssim 0.08$, with a slope β_t similar to the expected 1/2[51] (or $x_\epsilon = 2/7$,[27] $\alpha = -3/2$) characteristic of density fluctuations near a 3-state Potts tricritical point.

In a subsequent study of methane on graphite foam, Kim and Chan[52] used a.c. calorimetry because of its superior resolution over conventional adiabatic techniques for measuring the specific heat of broad, weak singularities. In this method[39,43], the sample is connected by a weak thermal link to a heat bath. One raises the sample temperature above that of the bath, apply an a.c. voltage to a heater on the sample,and finds thermal oscillations around this higher temperature. Over a range of system parameters, the heat capacity is inversely proportional to the rms magnitude of these oscillations and so can be easily extracted. (The proportionality constant over this range depends simply on the frequency and the peak power to the heater). In this system, at low enough T, there is a coexistence region between [2-d] vapor and [2-d] liquid, the phase boundary being determine by the peak of C in a fixed-θ scan. The top of the coexistence region in the T-θ phase diagram is flat and rather broad, but the critical coverage θ_c (as well as T_c) can be determined by noting that the boundary is nearly symmetric (about θ_c). Near θ_c plots of the specific heat (minus a smooth regular contribution) vs. reduced temperature are linear both above and below T_c for $0.025 \leq t \leq 0.01$, indicative of Ising-like logarithmic divergence. (The fact that $T > T_c$ and $T < T_c$ are co-plotted is consistent with the expected critical amplitude ratio of unity,[10] (cf. Section 2). Furthermore, the difference between the high or the low θ side and θ_c is expected to scale as t^{β}. Careful fits led to the estimate $\beta = 0.127 \pm 0.020$, in agreement with the Ising value of 1/8.

More recently Campbell and Bretz[39] used a.c. calorimetry to study the melting of the $(\sqrt{3}x\sqrt{3})R30^{\circ}$ phase of He on HOPG. The transition was expected to be 3-state Potts-like, but the measurements produced surprising results. Above T_c, α varied dramatically as a function of coverage: $\alpha = 0.37 \pm 0.1$ and $\alpha = 0.24 \pm 0.04$ for coverages presumably within ~ 0.2% of θ_m, the coverage having maximum T_c. (Here the low temperature phase has long-range order rather than coexistence.) However, as θ increases, α plummets, reaching -0.067 ± 0.10 about 2% above θ_m. This behavior is attributed to Fisher renormalization. On the other hand, by ~ 1/2% below θ_m, $\alpha = 0.48 \pm 0.05$. Berker[53] subsequently suggested this rise might be due to crossover to tricritical behavior, as there seems to be a tricritical point ~ 20% below θ_m. (Note $\alpha = 5/6$ for a 3-state Potts tricritical point approached in this way.) Below T_c, C exhibits logarithmic rather than power law behavior. The non-universal behavior is attributed to residual impurities. Indeed, the fits extend down only to t ~ 0.005, not as close as one would expect with the large L_s of HOPG.

4.3 Scattering Measurements.

Turning now to scattering experiments, we note first that 2-d Ising-like behavior was reported a decade ago using neutron scattering to study extra spots in antiferromagnetic materials with strong asymmetries producing 2-d behavior. As early as 1974 data from K_2CoF_4 was fit[54] beautifully to the Onsager magnetization formula. As another example, Birgeneau et al.[55] a few years later constructed log-log plots of peak width vs. reduced temperature for $0.009 \lesssim t \lesssim 0.15$ to estimate $\nu = 0.94$ (as compared to 1 for the Ising model). With little advancement in equipment, however, neutron scattering has made little real progress on this problem, and its weak scattering makes it not the optimal probe for surface studies.[56]

X-ray scattering with a rotating anode source was first applied by Horn et al.[57] to study Kr on ZYX graphite. The intensity of the peak induced by the $(\sqrt{3}x\sqrt{3})R30^{o}$ overlayer was fit to the form $t^{2\beta}$ for $t < 0.06$. Rather than trying a log-log plot and excluding the non-linear, finite-size smeared data near T_c, Horn et al. attempted to take the size-dependent shift of T_c[58] into account by convoluting $t^{2\beta}$ with a gaussian distribution of T_c's. (This method implicitly parameterizes an unknown correction-to-scaling function; no explicit justification has been given.) With this complicated fitting function, they obtained $\beta_{eff} = 0.09 \pm 0.03$, consistent with the expected 3-state Potts $\beta = 1/9$ (but also with the Ising and 4-state Pott models).

In a subsequent study of this system[59], the impressive improvement of L_i to $\sim 10^4$Å with a synchrotron source was used, along with the same method of analysis. Curiously, L_s also improved from 450Å to 2300Å, perhaps because only a small virgin portion of the sample near the center was illuminated. The smearing of the transition (the standard deviation of the gaussian) was accordingly reduced by nearly an order of magnitude, but the statistical error only by a factor of 2: $\beta_{eff} = 0.065 \pm 0.015$. This low value was attributed to crossover to 3-state Potts tricritical behavior ($\beta_M = 1/18$ [corresponding to $x_\sigma = 2/21$[27], $\alpha = 5/6$]), but later work[60] showed that the transition is in fact first order ($\beta = 0$). This example reinforces our earlier comment that reliance exclusively on measurements of the long-range order are risky and that study of critical scattering above T_c is more trustworthy. (Indeed, its absence, in retrospect, was an important clue that the interpretation was questionable.[61]) It also stimulates worries about the convolution fitting procedure. Nonetheless, we note that the behavior near weak first-order transitions can be virtually indistinguishable qualitatively from second-order, as we have illustrated explicitly in Monte Carlo simulations, similar to those in Section 3, of a p(2x2) overlayer on a honeycomb lattice (see Fig. 4 below) and of the closely-related 8-state Potts model.[62] The mere presence of what seems like critical scattering does not provide an "unequivocal signature"[63] of a continuous transition.[62]

More recently, most applications of synchrotron radiation have been to incommensuration[42,63] and especially to continuous (rather than lattice gas) 2-d melting[40,41], to check the predictions of the KTHNY theory[64]. For the latter, $\xi \propto \exp\,[\mathrm{const}/t^{\nu}]$, so that it is particularly important to have large L_i and L_s to see whether this form holds or if the transition is first order.

Campagna and co-workers used spin-polarized LEED (SPLEED) to study the magnetic critical behavior of Ni surfaces.[65] The difference between scattered intensity for incident electron spin polarization antiparallel and parallel to the [surface] magnetization of the target is measured. This magnetic exchange scattering is expected to measure the <u>surface</u> magnetization and so goes as $(T_{cs}-T)^{\beta_1}$ below the surface transition temperature T_{cs}, which is within 0.7% of the bulk T_c. Log-log plots of [normalized] exchange scattering vs. t are linear for $0.002 \lesssim t \lesssim 0.1$ (though with few points, having large fractional errors, below 0.01); the contrast with the previously discussed nonlinearities calculated for 2-d lattice gases is notable. For Ni (100), a square net, β_1^{eff} is 0.81 ± 0.02. For Ni(110), the top layer (a rectangular net) is contracted toward the bulk by 5-8%; nonetheless, $\beta_1^{\mathrm{eff}} = 0.79 \pm 0.02$, equal to the (100) value to within statistical uncertainly. For this "ordinary" surface transition, the scaling relation $\beta_1 + \gamma_1 = \beta + \gamma$ and ϵ -expansion calculations of γ_1 indicate $\beta_1 = 0.776 - 0.8$ for Ising [O(1)], 0.79-0.835 for XY [O(2)], and 0.81-0.88 for the expected Heisenberg [O(3)] model. Clearly [once again] it is difficult to distinguish universality classes based on just β_1. The leading correction to scaling is expected to be due to surface magnetic anisotropy; since this perturbation should be drastically different for the two surfaces, it cannot be the explanation of why β_1 is smaller than expected.[66]

The wide range of surface critical behavior[67] might also be seen in bimetallic alloys. For instance, LEED was used to look at the (100) face of Cu_3Au.[68] The (111) face of Cu-16% Aℓ has a $(\sqrt{3}\times\sqrt{3})R30^{\circ}$ structure,[69] suggestive of 3-state Potts, but since T_{cs} is close to T_c^{bulk}, the transition may be bulk-driven as was the magnetic transition on Ni.

The reconstruction of surfaces, i.e. the breaking of the 2-d space group of parallel layers in the bulk by primarily lateral distortions, should belong to 2-d universality classes (rather than being surface exponents of 3-d models); there is no underlying bulk transition. For example, LEED was recently used to steady the (1x2) reconstruction of the rectangular Au (110) surface.[46] Viewed on end, the unreconstructed surface can be described as an up and down (1/2-step) zig-zag of complete rows. In the reconstructed phase the width and depth of the zig-zag are effectively doubled, as though alternate rows in the top plane were removed; the details of course are more complicated. The reconstruction-induced spot was fit to a gaussian plus a lorentzian. Above T_c, log-log plots of intensity and width are shown for $0.04 \lesssim t \lesssim 0.2$, with exponents quoted as $\gamma = 1.75 \pm 0.03$ and $\nu = 1.02 \pm 0.02$. Below T_c, a log-log plot of the amplitude of the gaussian gives a surprisingly straight log-log plot for $0.006 < t < 0.1$, with $\beta_{eff} = 0.13 \pm 0.022$.

The quoted exponents are in excellent agreement with Ising values, but we believe the error bars are rather optimistic, with ± say 10% more consistent with our model calculations. Indeed, analyzing the exact Ising correlation length at the temperatures used in the fit yields an estimate of T_c 1/2% above the exact value and $\nu_{eff} = 0.89$. Various alterations of the fit of the depicted data points produced similar values of ν_{eff}. A subsequent integrated-intensity measurement[70] of a reportedly better gold surface, with T_c 8% higher, was fit for $0.004 \lesssim t \lesssim 0.035$ with equation (12) (but no A_1 term) to yield $\alpha = 0.02 \pm 0.05$, consistent with the Ising $\alpha = 0$ (logarithmic divergence) but not conclusive since linearity is expected in the finite-size rounded regime.

Finally we turn to the transitions of chemisorbed atoms on metal surfaces, the focus of work at Maryland. From the extensive tables of ordered states on surfaces of various symmetries,[71] one might expect these transitions to offer the most frequent examples of 2-d transitions. However, it turns out that with many systems there are peculiar complications. The melting of a c(2x2) overlayer on a square substrate is perhaps the simplest case. However, for the (100) faces of many bcc metals, especially W and Mo, the transitions at first attributed to the adatoms are actually reconstructions of the substrate.[38] Other problems also occur. For many cases, the transition turns out to be first order. For O on Ni(100), as the c(2x2) overlayer melts, the O atoms dissolve into the substrate, introducing uncontrollable bulk degrees of freedom.[72] For Cl on Ag (100), the gas atoms desorb [irreversibly] as the c(2x2) melts.[72] This behavior is related to a presumably extremely large nearest-neighbor repulsion.[73] In this case, however, one can observe a reversible transition to the ordered state by increasing θ at fixed T. (For O on Ni(100) there is a coexistence regime with a p(2x2) phase at coverages below the pure c(2x2) phase.) A log-log plot of [normalized] intensity vs. reduced coverage is linear (but over less than a decade of data). Since θ is varied, the exponent deduced from the slope is Fisher renormalized: $\beta/(1-\alpha)$ is estimated as 0.12 ± 0.03. This result is consistent with Ising exponents: $\beta = 1/8$ and $\alpha = 0$.

Another possible Ising realization is a p(2x1) on a rectangular lattice, i.e. alternate rows preferentially occupied in the ordered state. Wang and Lu[74] studied this structure in the case of O on W(112). Following Horn et al.[57] in convoluting T_c over a gaussian distribution, they analyzed $0.002 \lesssim t \lesssim 0.11$. Their effective exponents are quoted as $\beta = 0.13 \pm 0.01$, $\gamma = 1.79 \pm 0.14$, and $\nu = 1.09 \pm 0.11$.

To date none of the many examples of $(\sqrt{3}x\sqrt{3})R30^o$ chemisorbed states (on triangular nets) have been found to undergo continuous melting. For example, recent study of Al on Si(111) shows first-order behavior.[75] Systems worthy of further study include I on Cu(111)[76] and S on Pt(111).[77]

Many of the continuous transitions found in chemisorption systems belong to classes with c=1: the 4-state Potts model or the XY model with "cubic" anisotropy. As depicted in Fig. 4 the 3-fold sites of (111) faces of fcc (or hcp) metals two triangular sublattices, i.e. a honeycomb lattice with a binding energy

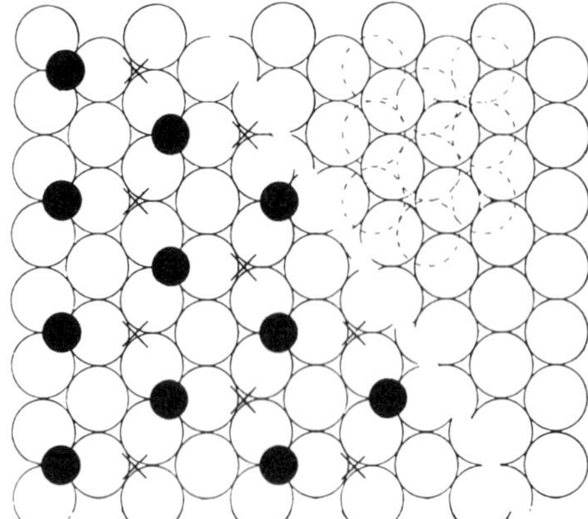

Fig. 4. Diagram to illustrate adsorption sites on (111) fcc surfaces. Open circles indicate atoms in the top layer. The symmetry between the two sublattices of 3-fold sites is broken by the second layer of atoms, illustrated by dashed circles in the upper right corner. A p(2x2) overlayer, e.g. O on Ni, is illustrated by the solid circles. For a graphitic (2x2), e.g. H on Ni, the sites marked by x's are also occupied.

difference (or staggered field) between the two kinds of 3-fold sites (one above second layer substrate atoms, the other above second-layer 3-fold sites and third-layer atoms.) In the case of H on Ni(111) the ordered state is a graphitic (2x2), i.e. a honeycomb with equal numbers of sites, one of four,on both sublattices occupied. The melting transition seems to be continuous;[78] it is expected to be in the 4-state Potts class. To the extent that the staggered field is negligible, there should also be a higher-temperature transition from a disordered low-θ regime to a (1x1) ordered state in which one of the two sublattices is preferentially occupied.[79] Unfortunately, it would be very difficult to measure such a transition since the ordered state introduces no new diffraction spots and the existing ("integer") spots are dominated by scattering from the substrate

For O on Ni(111), there are both p(2x2) and $(\sqrt{3}x\sqrt{3})R30^\circ$ ordered states, in which it is believed that sites in one of the two sublattices are preferentially occupied.[80] While the $\sqrt{3}x\sqrt{3}$ pattern disorders discontinuously, the p(2x2) -- first observed by Davisson and Germer[81] -- has a continuous transition expected to be in the 4-state Potts class. It was this venerable overlayer that was selected by our group for a first effort to extract critical exponents for a phase transition by chemisorbed atoms. Above T_c log-log plots of both [inverse] amplitude and width vs. were linear, albeit over less than a decade (0.015-0.08); the effective exponents were $\gamma = 1.9 \pm 0.2$ and $\nu = 0.94 \pm 0.10$ (with fits going only to 0.06). Below T_c there was relatively little data, since it was not till after deconvolution of the instrument response that we found T_c was lower than it had seemed. With T_c fixed at the value obtained from $T > T_c$ fits, we estimated $\beta = 0.14 \pm 0.02$ (with T_c variable, T_c rose 1/2% and β increased to 0.16). These values are much closer to Ising than to the expected 4-state Potts results. Our checks showed that the logarithmic corrections of the latter can account partially, but far from fully, for these effective values. Reconstituting the data to show what would be a seen in an integrated intensity measurement showed no apparent anomaly near T_c, consistent with Ising $\alpha = 0$.[82] While a variety of explanations have been suggested, it may well just be that random defects smeared (destroyed) the transition.

Simultaneous but cruder work treated H on W(110),[83] a centered rectangular net of sites. Both the p(2x1) [near $\theta = 1/2$] and the (2x2) [near $\theta \sim 3/4$] ordered states are expected to have disordering transitions in the XY-with-cubic-anisotropy class. Since the exponents are non-universal, it is hard to calibrate results. Only the intensity below T_c was analyzed, using a gaussian distribution of T_c's[57]. For $0.05 < t < 0.2$, linear log-log plots were obtained, and the exponent β was determined to be 0.13 ± 0.04 and 0.25 ± 0.07, respectively, for the two transition.

This procedure was also followed in a recent analysis of the reconstruction of the (100) face of W.[84] A log-log plot of intensity vs. t for this transition, expected also to be in the XY with cubic anisotropy class, was linear only for $0.035\ t \leq 0.08$, and the deduced $\beta_{eff} = 0.144 \pm 0.04$.

As a final example, we note the intriguing system of Se on Ni(100).[85] This system has both a c(2x2) and a lower-coverage p(2x2) ordered phase and might offer the first realization of the Ashkin-Teller model.[86] That venerable model involves two Ising models on the same lattice, coupled by a quadratic term and has three phases: 1) both Ising lattices ordered [corresponding to p(2x2)], 2) a "polarized" regime with each Ising lattice disordered but with non-zero expectation value of the product of spins from both [corresponding to c(2x2)], 3) disordered. The three phases meet at a 4-state Potts point. The transition between the ordered and disordered phases is XY with cubic anisotropy, while the lines between the polarized phase and either of the others are both Ising. The critical properties of the lattice gas have yet to be measured.

V. CLOSING COMMENTS

We have presented a survey of experimental studies of 2-d lattice gas phase transitions along with a suggestion based on Monte Carlo calculations of what one might expect to learn. At present, one is pressed to obtain the leading exponents, let alone some estimate of corrections to scaling. There are other aspects that theorists would like to have measured but are beyond current capabilities.

Most intriguing is the idea of testing the finite size scaling formula for the free energy by measuring the specific heat of vicinal (stepped, with wide terraces) surfaces. Obviously there will not be periodic boundary conditions, so equation (20) becomes[35]

$$f = f_{bulk} + 2f_{edge}/L - \pi\, c/24L^2, \qquad (21)$$

where L now is the terrace width. Free boundary conditions are also unphysical. While some sort of fixed boundary condition might apply, it is noteworthy that the binding energy at the top edge of a step is different from that at the bottom. One could at least in principle get some idea of these energies by studying distributions of adatoms. Individual atoms can now be imaged experimentally using the scanning tunneling microscope[87] (well publicized due to the recent Nobel Prize) or the field ion microscope, in which He atoms ionize at a positively charged and then are accelerated to a large hemispherical negatively-charged screen, thereby providing a magnified image of the tip.[88]

Another wish would be to study the 3-site correlation function, for which conformal invariance makes specific predictions.[89] It is unlikely that such multiple scattering contributions, alluded to in Section 2, can be separated out in adequate detail. It is perhaps worth stating here that multiple scattering is, however, not as problematic for scattering experiments as once thought.[90] If integrated measurements are made, they are clearly not a difficulty since multi-site correlation functions contain the same energy-like anomaly at T_c as pair functions.[22] In the diffraction limit, multiple scattering will just change the amplitudes of existing corrections to scaling, for "extra" spots induced by the ordering.[90]

There is some hope that with high-resolution LEED, the scattering line shape near T_c can be fit to the predictions of conformal invariance. Work in this direction is well underway.[91]

A major impediment to experimental progress is the notion that critical phenomena are too hard to measure, too esoteric to understand, too passé to be exciting, or too abstract to be funded. It is thus important for experts in conformal invariance interact meaningfully with their experimental colleagues.

ACKNOWLEDGEMENTS

This work and my research has been funded by the US Department of Energy under grant DE-FG05-84ER45071. This research would have been impossible without the stimulating, energetic collaboration of N.C. Bartelt and L.D. Roelofs. I thank the former for many helpful comments on this presentation. I also thank P. Kleban for turning our attention to conformal invariance and apprising us of important developments. Ongoing fruitful interactions with the experimental group studying 2-d phase transitions at Maryland, led by R.L. Park and E.D. Williams, is happily acknowledged. Finally, I thank T.Y. Contee for ably typing the manuscript as I stood over her shoulder and V. Rittenberg stood over mine.

REFERENCES

1. For a review, see Einstein, T.L. in Chemistry and Physics of Solid Surface II, Vanselow, R., ed. (CRC, Boca Raton, 1979), 181.

2. Huse, D.A. Phys. Rev. B 29, 5031 (1984).

3. Bartelt, N.C. Ph.D. dissertation, University of Maryland, 1986 (unpublished).

4. Webb, M.B. and Lagally, M.G., Solid State Phys. 28, 305 (1973); Pendry, J.B., Low-Energy Electron Diffraction (Academic New York, 1974); Marcus, P.M. and Jona, F., Determination of Surface Structure by LEED (Plenum, New York, 1984). For a plot of the "universal curve of electron mean free path vs. energy, see e.g. Seah, M.P. and Dench, W.A., Surface Interface Annal. 1, 2 (1979)); Powell, C.J., ibid. 3, 94 (1981).

5. Engel, T., in Chemistry and Physics of Solid Surfaces V, Vanselow, R. and Howe, R., eds (Springer, Berlin, 1984), 205.

6. Kantor, D.F. and Lynn, K.G., J. Vac. Sci. Technol. A 2, 916 (1984).

7. Henzler, M., Surface Sci 152/153, 963 (1985); Lagally, M.G. Appl. Surface Sci. 13, 260 (1982).

8. Tracy, C.A. and McCoy, B.M., Phys. Rev. B 12, 368 (1975).

9. An excellent general reference in critical phenomena is Pfeuty, P. and Toulouse, G., Introduction to the Renormalization Group and Critical Phenomena (Wiley, New York, 1977); see also Fisher, M.E., in Collective Properties of Physical Systems, Lundqvist, B. and Lundqvist, S., eds. (Proc 24th Nobel Symposium, Academic, New York, 1973), 16.

10. Kaufman, M. and Andelman, D. Phys. Rev. B 29, 4010 (1984).

11. For a more detailed discussion, with a figure, see Einstein, T.L., in Chemistry and Physics of Solid Surfaces IV, Vanselow, R. and Howe, R. eds. (Springer, Berlin, 1982), 251, which also amplifies some of the other topics of this section, but is somewhat outdated.

12. Fisher, M.E., Phys. Rev. 176, 257 (1968).

13. Lawrie, I.D. and Sarbach, S., in Phase Transitions and Critical Phenomena, Domb, C. and Lebowitz, J.L., eds., vol. 9 (Academic London, 1984), Chap 1.

14. Aharony, A., in Critical Phenomena, Hahne, F.J.W., ed. (Springer, Berlin, 1983), 209.

15. Berker, A.N., Phys. Rev. B 12, 2752 (1975); Caflisch, R. G. and Berker, A. N., Phys. Rev. B 29, 1279 (1984).

16. Domany, E., Schick, M., Walker, J.S., and Griffiths, R.B., Phys. Rev. B 18, 2209 (1978).

17. Schick, M., Prog. Surf. Sci. 11, 245 (1981).

18. Bartelt, N.C., Einstein, T.L., and Roelofs, L.D., Phys. Rev. B. accepted for publication.

19. Alexander, S., Phys. Lett. 54A, 353 (1975).

20. Bartelt, N.C., Einstein, T.L., and Roelofs, L.D., Phys. Rev. B 35, xxx (1987).

21. Fisher, M.E., and Langer, J.S., Phys. Rev. Lett. 20, 665 (1978); see also Ferer, M., Moore, M.A., and Wortis, M., ibid. 22, 1382 (1969).

22. Bartelt, N.C., Einstein, T.L., Roelofs, L.D., Phys. Rev. B 32, 2993 (1985).

23. For a review of results on Potts models see Wu, F.Y., Rev. Mod. Phys. 54, 235 (1982).

24. Bartelt, N.C., and Einstein, T.L., J. Phys. A 19, 1429 (1986). The appendix shows how the structure factor can be computed rather simply from transfer matrices.

25. Kleban, P., Akinci, G., Hentschke, R., and Brownstein, K.R., J. Phys. A 19, 439 (1986).

26. Nienhuis, B., J. Phys. A 15, 199 (1982).

27. Friedan, D., Qiu, Z, and Shenker, S. Phys. Rev. Lett. 52, 1575 (1984); in Vertex Operators in Mathematics and Physics, Lepowsky, J. Mandelstam, S., and Singer, I.M., eds. (Springer, New York, 1984), 419.

28. The ratio square is Σ_2^+/Σ_2^- of equation. (2.14) of Ref. 8. see also Barouch, E., McCoy, B.M., and Wu, T.T., Phys. Rev. Lett. 31, 1409 (1973).

29. Huse, D.A., and Fisher, M.E., Phys. Rev. Lett. 49, 793 (1982); Phys. Rev. B 29, 239 (1984).

30. den Nijs, M.P.M., J. Phys. A 17, L295 (1984).

31. Saito, Y., Phys. Rev. B 24, 6652 (1981).

32. Bartelt, N.C., Einstein, T.L., and Roelofs, L.D., Phys. Rev. B 34, 1616 (1986).

33. Hamer, C.J., Phys. A 15, L675 (1982); 16, 3085 (1983).

34. Baxter, R.J., J. Phys. A 13, L61 (1980); J. Stat Phys. 26, 427 (1981); Baxter, R.J., and Pearce, P.A., J. Phys. A 16, 2239 (1983); Huse, D.A., Phys. Rev. Lett. 49, 1121 (1982).

35. Blöte, H.W.J., Cardy, J.L., and Nightingale, M.P., Phys. Rev. Lett. 56, 742 (1986); Affleck, I., ibid. 56, 746 (1986).

36. Cf., e.g. Zhao, L.-H., Lu, T.-M., and Lagally, M.G., Appl. Surface Sci. 11/12, 634 (1982); Pimbley, J.M. and Lu, T.-M., J. Vac. Sci. Technol. A 2, 457 (1984).

37. For a review, see Kleban, P., in Chemistry and Physics of Solid Surfaces V, Vanselow, R., and Howe, R., eds. (Springer, Berlin 1984), 339.

38. For a review of reconstruction, see Estrup, P.J., in Chemistry and Physics of Solid Surfaces V, Vanselow, R., and Howe, R., eds. (Springer, Berlin, 1984), 205.

39. Campbell, J.H., and Bretz, M., Phys. Rev. B 32, 2861 (1985).

40. Nagler, S.E., Horen, P.M., Rosenbaum, T.F., Birgeneau, R.J., Sutton, M., Mochrie, S.G.J., Moncton, D.E., and Clarke, R., Phys. Rev. B 32, 7373 (1985).

41. Heiney, P.A., Stephens, P.W., Birgeneau, R.J., Horn, P.M., and Moncton, D.E., Phys. Rev. B 28, 6416 (1983); Dimon, P., Horn, P.M., Sutton, M., Birgeneau, R.J., and Moncton, D.E., Phys. Rev. B 31, 437 (1985).

42. Stephens, P.W., Heiney, P.A., Birgeneau, R.J., Horn, P.M., Moncton, D.E., and Brown, G.S., Phys. Rev. B 29, 3512 (1984).

43. Miner, K.D., Chan, M.H.W., and Migone, A.D., Phys. Rev. Lett. 51, 1465 (1983).

44. Lagally, M.G., and Martin, J.A., Rev. Sci. Instrum 54, 1273 (1983); Scheithauer, U., Meyer, G., and Henzler, M., Surface Sci. (Proc. ECOSS-8), xxx (1987).

45. Foster, M.S., Campuzano, J.C., Willis, R.F., and Dupuy, J.C., J. Microscopy 140, 395 (1985).

46. Campuzano, J.C., Foster, M.S., Jennings, G., Willis, R.F., and Unertl, W., Phys. Rev. Lett. 54, 2684 (1985); Campuzano,, J.C., Jennings, G., and Willis, R.F., Surface Sci. 162, 484 (1985).

47. Toennies, J.P., private communication, 1986.

48. Tejwani, M., Ferreira, O., and Vilches, O.E., Phys. Rev. Lett. 44, 152 (1980); also Bretz, M., Phys. Rev. Lett. 38, 501 (1977).

49. Cf. footnote 40 of Ref. 22.

50. Migone, A.D., Chan, M.H.W., Niskanen, K.J., and Griffiths, R.B., J. Phys. C 16, L1115 (1983).

51. Ostlund, S., Berker, A.N., Phys. Rev. B 21, 5410 (1980).

52. Kim, H.K., Chan, M.H.W., Phys. Rev. Lett. 53, 170 (1984).

53. Berker, A.N., comment at March 1985 American Physical Society Meeting, Baltimore.

54. Ikeda, H., and Hirakawa, K., Solid State Comm. 14, 529 (1974).

55. Cowley, R.A., Birgeneau, R.J., and Shirane, G. in Strongly Fluctuating Condensed Matter Systems, Riste, T., ed. (Plenum, New York, 1979); Birgeneau, R.J., Als-Nielsen, J., and Shirane, G., Phys. Rev. B 16, 280 (1977).

56. See, however, "Neutron Scattering Applied to Surfaces," Thomas, R.K., in Emission and Scattering Techniques, Day, P., ed. (Reidel, D., Dordrecht, Holland, 1981), 251.

57. Horn, P.M., Birgeneau, R.J., Heiney,, P., and Hammonds, E.M., Phys. Rev. Lett. 41, 961 (1978).

58. E.g. Ferdinand, A.E., and Fisher, M.E., Phys. Rev. 185, 832 (1969).

59. Birgeneau, R.J., Brown, G.S., Horn, P.M., Moncton, D.E., and Stephens, P.W., J. Phys. C 14, L49 (1981).

60. Specht, E.D., Sutton, M., Birgeneau, R.J., Moncton, D.E., and Horn, P.M., Phys. Rev. B 30, 1589 (1984).

61. Stephens, P.W., Heiney, P., private communications.

62. Bartelt, N.C., Einstein, T.L., and Roelofs, L.D., submitted to Phys. Rev. B.

63. McRae, E.G., and Malic, E.G., Surface Sci. 161, 25 (1985).

64. Stephens, P.W., Goldman, A.I., Heiney, P.A., and Bancel, P.A., Phys. Rev. B 33, 655 (1986).

65. Kosterlitz, J.M., and Thouless, D.J., J. Phys. C 5, L124 (1972); 6, 1181 (1973); Halperin, B.I., and Nelson, D.R., Phys. Rev. Lett. 41, 121 (1978); 41, 519 (E) (1978); Nelson, D.R., and Halperin, B.I., Phys. Rev. B 18, 2318 (1978); 19, 2457 (1979); Young, A.P., Phys. Rev. B 19, 1855 (1979).

66. Alvarado, S.F., Campagna, M., Ciccacci, F., and Hopster, H., J. Appl. Phys. 53, 7920 (1982); also Phys. Rev. Lett. 48, 51 (1982).

67. For a review, see Binder, K., in Phase Transitions and Critical Phenomena, Domb, C., and Lebowitz, J.L., eds. vol. 8 (Academic, London, 1983), 1.

68. McRae, E.G., and Malic, R.A., Surface Sci. 148, 551 (1984).

69. Baird, R.J., Ogletree, D.F., VanHove, M.A., and Somorjai, G.A., Surface Sci. 165, 345 (1986).

70. Clark, D.E., Unertl, W.N., and Kleban, P.H., Phys. Rev. B 34, 4379 (1986).

71. Somorjai, G.A., Chemistry in Two Dimensions: Surfaces (Cornell U. Press, Ithaca, 1981); Somorjai, G.A. and Szalkowski, F.Z., J. Chem. Phys. 54, 389 (1971).

72. Taylor, D.E., and Park, R.L., Surface Sci. 125, L73 (1983).

73. Taylor, D.E., Williams, E.D., Park, R.L., Bartelt, N.C., and Einstein, T.L., Phys. Rev. B 32, 4653 (1985).

74. Wang, G.-C., and Lu, T.-M., Phys. Rev. B 31, 5918 (1985).

75. Hwang, R.Q., Williams, E.D., and Park, R.L., Bull Am. Phys. Soc. 32, xxx (1987).

76. DiCenzo, S.B., Wertheim, G.K., and Buchanan, D.N.E., Surface Sci. 121, 411 (1982).

77. Auer, M., Leonhard, H., and Hayek, K., Appl. Surface Sci. 17, 70 (1983).

78. Christmann, K., Behm, R.J., Ertl, G., VanHove, M.A., and Weinberg, W.H., J. Chem. Phys. 70, 4168 (1979).

79. Roelofs, L.D., Einstein, T.L., Bartelt, N.C., and Shore, J.D., Surface Sci. 176, 295 (1986).

80. Roelofs, L.D., Kortan, A.R., Einstein, T.L., and Park, R.L., Phys. Rev. Lett. 46, 1465 (1981).

81. Davisson, C. and Germer, L.H., Phys. Rev. 30, 705 (1927).

82. Bartelt, N.C., Einstein, T.L., and Roelofs, L.D., in The Structure of Surfaces, VanHove, M.A., and Tong, S.Y., eds. (Springer, Berlin, 1985), 357

83. Lyuksyutov, I.F., and Fedorus, A.G., Sov. Phys. JETP 53, 1317 (1981) [Zh. Eksp. Teor. Fiz. 80, 2511 (1981)].

84. Wendelken, J.F., and Wang, G.-C., Phys. Rev. B 32, 7542 (1985).

85. Bak, P., Kleban, P., Unertl, W.N., Ochab, J., Akinci, G., Bartelt, N.C., and Einstein, T.L., Phys. Rev. Lett. 54, 1539 (1985).

86. Ashkin, J., and Teller, E., Phys. Rev. 64, 178 (1943).

87. Behm, R.J., and Hösler, W., in Chemistry and Physics of Solid Surfaces VI, Vanselow, R., and Howe, R., eds. (Springer, Berlin, 1986).

88. Müller, E.W., in Chemistry and Physics of Solid Surfaces - I, Vanselow, R., and Tong, S.Y., eds. (CRC, Cleveland, 1977), 1.

89. Cardy, J.L., in Phase Transitions and Critical Phenomena, Domb, C. and Lebowitz, J.L., eds. vol. 11 (Academic, London, 1987).

90. Bartelt, N.C., Einstein, T.L., and Roelofs, L.D., Phys. Rev. Lett. 56, 2881 (1986).

91. Kleban, P., Hentschke, R., and Campuzano, J.C., preprint.

The Vertex Operator Construction for Non-simply Laced Kac-Moody Algebras

David I. Olive
Blackett Laboratory, Imperial College, London SW7 2BZ.
Dept. de Physique Theorique, University of Geneva.

When g is a simply-laced simple Lie algebra i.e. A_r (su(r+1)), D_r (so(2r+1)), E_6, E_7, or E_8, the vertex operator construction[1,2] of the level one representations of the affine untwisted Kac-Moody algebra $\hat{g}$ obtained by affinising g involves the emission vertex for the tachyon states of the bosonic string theory[3]. The vertices for all the states of this theory generate a Lorentzian algebra of rank 26 based on the unique self-dual even lattice in 26 Lorentzian dimensions which seems closely related to the bosonic string theory[4,5].

Here we outline the construction of a level one representation of $\hat{g}$ when g is not simply laced i.e. is of type B_r (so(2r+1)), C_r (sp(r)), F_4 or G_2, due independently to Goddard, Nahm, Olive and Schwimmer[6] and to Bernard and Thierry-Mieg[7]. It will turn out that this new construction relates to the fermionic string theory of Ramond[8], Neveu and Schwarz[9], at least when the ratios of the squares of the lengths of the long roots to the short roots, denoted $(L/S)^2$, equals 2 i.e. for B_r, C_r and F_4 but not G_2 which we henceforth exclude for simplicity of presentation. The reason is that this string theory possesses two tachyons, with mass squared -2 and -1 respectively, whose vertex operators are both used. Actually the first tachyon decouples in the physical theory even though its Regge recurrences do not (and the second, the "dual pion", decouples in the space-time supersymmetric version of the theory[10]).

It has recently been understood[11,12] that a powerful

tool in the construction of highest weight representations of $\hat{g}$ (i.e. ones built on a ground state) is provided by the Sugawara construction of the Virasoro algebra[13] generators as bilinears in the generators of $\hat{g}$. Extremely useful information is provided by the c-number of this Virasoro algebra, which takes the following value in a representation of $\hat{g}$ with level x

$$c = \frac{x \dim g}{x + \tilde{h}(g)} = \text{rank } g + \frac{n_L(x-1) + n_S(x-(S/L)^2)}{x + \tilde{h}(g)} \tag{1}$$

where $\tilde{h}(g)$ is the dual Coxeter number of g, an integer related to the quadratic Casimir in the adjoint representation, and n_L and n_S are the number of long roots and short roots of g respectively.

As x is an integer larger or equal to 1, we see that c exceeds the rank of g unless x=1 and $(L/S)^2 = 1$ i.e. g is simply laced, when c and rank g coincide. For an abelian group, $\tilde{h}$ vanishes and so by (1) c equals its rank or, what is the same thing, its dimension. But G, the group obtained by exponentiating g, contains an abelian subgroup of the same rank as g obtained by exponentiating the Cartan subalgebra t. Thus $\hat{g}$ and $\hat{t}$ yield the same c value and this suggests that $\hat{g}$ can be constructed from $\hat{t}$ in this case. This is what the vertex operator construction does as seen below. When $(L/S)^2 = 2$ rather than 1, (1) yields instead

$$c = \text{rank } g + \frac{n(n+1)}{n+3} \tag{2}$$

where n is the number of short simple roots of g, 1 for B_r, r-1 for C_r and 2 for F_4. This value (2) for c is neither an integer nor a half integer unless n = 1, 3 or 9. This means that the Virasoro generators relevant to level 1 representations of g with g = B_r, C_r and F_4 cannot, in general, be constructed from free fields, bosonic or even fermionic, but must involve fields which interact with each other in some sense. This is the essential new complication of interest.

When g is not simply-laced, its long roots form the root system of a subalgebra of g, denoted g_L, of the same rank. Since g_L is automatically simply-laced we can take its roots to have length $\sqrt{2}$ and construct a level 1 representation of $\hat{g}_L$ by the usual Frenkel, Kac and Segal construction[1,2]

$$\sum_{m\varepsilon Z} z^{-m} E^{\alpha}_m = z{:}e^{i\alpha.Q(z)}{:}c_\alpha ;\ \alpha^2 = 2 \tag{3}$$

$$\sum_{m\varepsilon Z} z^{-m} H^i_m = iz\frac{dQ^i}{dz}, \tag{4}$$

where $Q^i(z)$ is the Fubini-Veneziano coordinate vector[14] and c_α the Klein transformation factor[15]. See chapter 6 of my recent review article with Peter Goddard for notation and details[12]. To construct $\hat{g}$ given $\hat{g}_L$ it only remains to express the step operators associated with the short roots of g (which have unit length):

$$\sum_{m\varepsilon Z} z^{-m} E^{e}_m = \sqrt{z}{:}e^{ie.Q(z)}{:}c_e\Psi_\Omega(z) ;\ e^2 = 1,\ e\ \varepsilon\Omega , \tag{5}$$

where Ψ_Ω are the new fields accounting for the term n(n+1)/(n+3) in (2) and are the same for each e in a given subset Ω of the short roots. Ω will be defined below.

By dimensional arguments all the generators of $\hat{g}$ have conformal weight 1, as is guaranteed by the ansätze (3) and (4) since $z^{x^2/2}{:}e^{ix.Q(z)}{:}$ has conformal weight $x^2/2$. Comparing with (5) we see that $\Psi_\Omega(z)$ must therefore carry the missing conformal weight 1/2, indicating that it is actually a fermionic field. Its fermionic nature is expressed by the operator product expansion

$$\Psi_\Omega(z)\Psi_\Omega(\zeta) = {}^{\circ}_{\circ}\Psi_\Omega(z)\Psi_\Omega(\zeta)^{\circ}_{\circ} + \Delta(z,\zeta), \quad |z|>|\zeta| , \tag{6}$$

where the fermionic normal ordered product denoted by open dots is antisymmetric in z and ζ and regular at $z = \zeta$, actually vanishing there, while the c-number contraction, Δ, has a pole there with residue z.

The check of the commutators of (4) with (5) is straightforward as is that of (3) with (5) provided $e+\alpha\epsilon\Omega$ whenever $e\epsilon\Omega$ and $e.\alpha=-1$. Since $e+\alpha$ is the Weyl reflection of e in α we deduce that the sets Ω are actually the orbits of the short roots of g under the action of the Weyl group of g_L. This is generated by reflections in the long roots of g. These orbits are all isomorphic and have the property that $-\Omega=\Omega$ from which we deduce that Ψ_Ω is a real fermion field (as already assumed in (6)). The check of the commutators of (5) with itself are again straight forward using (6). The remaining commutators, between E^e_m and E^f_n, where e and f belong to distinct orbits, gives the desired results if we make the following additional assumptions about Ψ_Ω ,

$$\Psi_\Omega(z)\Psi_{\Omega'}(\zeta) = \Sigma(\Omega,\Omega')\Psi_{\Omega'}(\zeta)\Psi_\Omega(z) \tag{7}$$

is regular, where $\Sigma=\pm1$ and the points of Ω and Ω' are all mutually perpendicular. The remaining possibility is that the points of Ω and Ω' have scalar products $\pm1/2$. Then we must assume

$$\Psi_{\Omega'}(z)\Psi_{\Omega_2}(\zeta) = (z-\zeta)^{-1/2}\{\eta_+(\Omega_1,\Omega_2)R^+_{\Omega_1,\Omega_2}(z,\zeta)$$

$$+ (z-\zeta)\eta_-(\Omega_1,\Omega_2)R^-_{\Omega_1\Omega_2}(z,\zeta) , \quad |z|>|\zeta| \tag{8}$$

where $R^{\pm}_{\Omega_1,\Omega_2}(z,\zeta)$ are symmetric under the simultaneous interchange of Ω_1 with Ω_2 and z with ζ and

$$R_{\Omega_1,\Omega_2}(z,z) = \sqrt{z}\;\Psi_{\Omega_3}(z)/\sqrt{2} \qquad (9)$$

It is equations (8) and (9) which express the fact that the fermion fields for different orbits are not independent and hence interact in some sense.

In checking the required commutators one needs to assume that the "cocycles" Σ and $\eta_{\pm}$ have certain properties. Thus one ought to check that such quantities can indeed be constructed as can the fermion fields Ψ_{Ω}. This is the most intricate part of the analysis and details can be found in our paper, but a key point concerns the relative structure of the root lattices of g and g_L, $\Lambda_R(g)$ and $\Lambda_R(g_L)$ respectively. They satisfy the following properties (if Λ and Λ_o denote $\Lambda_R(g)$ and $\Lambda_R(g_L)$ respectively)

$$\Lambda_o \text{ and } \sqrt{2}\Lambda \text{ are even integral lattices} \qquad (10a)$$

$$\Lambda_o \subset \Lambda \subset \Lambda_o^{*} \qquad (10b)$$

$$\Lambda/\Lambda_o = (\mathbb{Z}_2)^n. \qquad (10c)$$

These are the properties used, and as a consequence the work applies to other solutions to (10). The aforementioned orbits are the sets of points of unit length in each distinct coset Λ/Λ_o. $\mathbb{Z}_2$ is the cyclic group with two elements and the power to which it occurs in (10c), namely n, is the number of short simple roots of g, as in (2). As well as specifying the multiplication law for cosets, (10c) also specifies the multiplication law for the orbits and hence for the associated fermion fields as in (8) and (9).

There are other solutions to (10), most notably

$$\Lambda = \Lambda_R(su(n))/\sqrt{2}, \; \Lambda_o = \sqrt{2}\Lambda_R(su(n)) \qquad (11)$$

where Λ_R is normalised so that the roots have length $\sqrt{2}$ and hence unit length in Λ. Thus our construction can be applied to su(n) too by treating its roots as short roots rather than as long roots as in the Frenkel, Kac and Segal construction[1,2].

The interest arises when we count the number of short roots in an orbit. It is 2 for su(n) as given by (11), 4 for C_{n+1} and 8 for F_4 i.e. the number of complex numbers, quaternions and octonions respectively. This is not accidental, equations (8) and (9) provide a multiplication law on the orbits which can be identified with these division algebras[6,16]. B_n has only one orbit and so is irrelevant from this point of view. When n matches and certain other matching conditions are met it is possible to define a multiplication operation between two different algebras and obtain a third algebra which is simply laced. This leads to Freudenthal's magic square and generalisations[6,17]. We regard this structure as particularly interesting and likely to be of importance in superstring theory.

Another feature[6] concerns the fact that the Lie algebras we have mentioned can be obtained from simply laced algebras of higher rank by "folding" their Dynkin diagram according to a $\mathbb{Z}_2$ symmetry:

$D_{r+1} \Rightarrow B_r$

$A_{2r-1} \Rightarrow C_r$

$E_6 \Rightarrow F_4$

$(A_r)^2 \Rightarrow A_r$

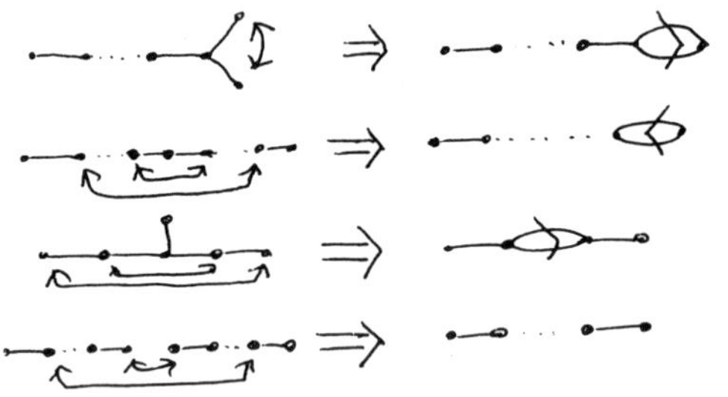

The constructions (3), (4) and (5) can be understood in terms of the simply-laced construction (3) and (4) alone on the basis of extending the above symmetry τ to the whole algebra and considering the fixed points. If the two roots α and $\tau(\alpha)$ of the simply-laced algebra are unequal they satisfy $\alpha.\tau(\alpha) = 0$. Then $(\alpha+\tau(\alpha))/2$ is a unit vector and this can be used to illustrate how the fermion fields in (5) arise out of the folding:

$$z{:}e^{i\alpha.Q(z)}{:} + z{:}e^{i\tau(\alpha).Q(z)}{:} =$$

$$\sqrt{2z}{:}e^{i(\alpha+\tau(\alpha))/2.Q(z)}{:}\Psi_{\Omega}(z)$$

where $\Psi_{\Omega}(z) = \sqrt{z}{:}\cos(\alpha-\tau(\alpha))/2.Q(z){:}/\sqrt{2}$.

Apart from the occurrence of octonions as already mentioned the interest of this for superstring theory is the resemblance of the two vertices used in (3) and (5) to those for emitting tachyons with mass squared -2 and -1 respectively in the Ramond-Neveu-Schwarz model. The above folding procedure indicates how it might be possible to obtain this fermionic string theory from a purely bosonic theory in higher dimensions.

We believe that these ideas are capable of further development and could greatly enhance our understanding of the mathematical structure of the superstring theory which has long been the leading candidate for the unified theory of particle interactions.

Further explanation and details will be given by Peter Goddard in his accompanying article.

REFERENCES

1) I.B. Frenkel and V.G. Kac, Inv. Math. 62 (1980) 23.
2) G. Segal. Comm. Math. Phys. 80 (1981) 301.
3) M. Jacob (Editor), "Dual Theory" (North Holland, 1974).
4) P. Goddard and D. Olive, "Vertex Operators in Mathematics and Physics", MSRI publication 3 (Springer, 1984), p51.
5) I.B. Frenkel, "Proceedings of meeting on Applications of Group Theory in Physics and Mathematical Physics", Chicago 1982, Lectures in Applied Maths. 21 (1985) 355.

6) P. Goddard, W. Nahm, D. Olive and A. Schwimmer, "Vertex Operators for Non-simply-laced Algebras" Commun. Math. Phys. 107 (1986) 179.
7) D. Bernard and J. Thierry-Mieg, Meudon preprint.
8) P. Ramond, Phys. Rev. D3 (1971) 2415.
9) A. Neveu and J.H. Schwarz, Nucl. Phys. B31 (1971) 86; Phys. Rev. D4 (1971) 1109.
10) F. Gliozzi, D. Olive and J. Scherk, Phys. Lett. 65B (1976) 282; Nucl. Phys. B122 (1977) 253.
11) D. Olive, "Kac-Moody and Virasoro Algebras in Local Quantum Physics" Imperial/TP/84-85/33, (1985 Erice lectures).
P. Goddard and D. Olive, "Unified String Theories" ed. by M. Green and D. Gross (World Scientific, Singapore, 1986) p. 214.
12) P. Goddard and D. Olive "Kac-Moody and Virasoro Algebras in Relation to Quantum Physics" DAMTP preprint 86-5, June 1986.
13) H. Sugawara, Phys. Rev. 170 (1968) 1659.
14) S. Fubini and G. Veneziano, Nuovo Cim. 67A (1970) 29.
15) O. Klein, J. Phys. Radium 9 (1938) 1.
16) P. Goddard, D. Olive, W. Nahm, H. Ruegg and A. Schwimmer, to appear.
17) P. Goddard, D. Olive and A. Schwimmer, Phys. Lett. 157B (1985) 393.

REPRESENTATION THEORY OF THE VIRASORO AND SUPER-VIRASORO ALGEBRAS: IRREDUCIBLE CHARACTERS†

Alvany Rocha-Caridi

Department of Mathematics, Baruch College, C.U.N.Y.
17 Lexington Avenue, New York, N.Y. 10010
U.S.A.

ABSTRACT

An analytic vector (the Shapovalov element), giving embeddings of Verma modules over the Virasoro algebra, is constructed. Using the Shapovalov element, a character sum formula for the quotient of two Verma modules is obtained. The formula leads to an alternative way of deriving formulas for the irreducible characters. This is carried out in the case of the minimal conformal invariant theories representations. In particular, a new derivation of the irreducible characters of the unitary discrete series is presented. The case of the Super-Virasoro algebra is also treated.

1. INTRODUCTION

Since the important discovery by Friedan, Qiu and Shenker (F-Q-S) of the discrete series of representations of the Virasoro algebra[1,2], this algebra has played in increasing role in two-dimensional quantum field theory. The F-Q-S discrete series is a family of irreducible vacuum vector representations parametrized by a discretely infinite set of values of h (the energy) and c

† 10th Johns Hopkings Workshop on Current Problems in Particle Theory, Bad Honnef, Federal Republic of Germany, September 1 - 3, 1986, ed. by V. Rittenberg, World Scientific.

(the central charge), whose complement in the strip $0 \leq c < 1$ corresponds to non-unitarizable representations[1,2,3] (see also [4]). This set of parameters is a subset of the one defining the minimal conformal invariant theories of Belavin, Polyakov and Zamolodchikov (B-P-Z)[5]. It has now been established that the discrete series representations are unitarizable[6,7] and, therefore, that they give a catalog of the conformally invariant models in statistical mechanics[1,2].

The relationship between the irreducible characters and the description of the subrepresentations of Verma modules was established by the author[8] (see also [9] for the case $c = 0$). In particular, explicit formulas for the irreducible characters of the unitary discrete series were obtained[8] using the subrepresentation structure announced by Feigin and Fuchs[10]. A subsequent publication of a sketch of the proofs of the results announced in [10][11] also discusses the relationship with irreducible characters. Among the applications of the characters of the discrete series are the proof of their unitarity[6,7] and the study of modular invariance in conformally invariant theories[12,13,14,15,16,17].

In this paper I shall present a self-contained derivation of the characters of the B-P-Z minimal theories representations. In particular, I shall give a proof of the unitary series character formulas without using the results of [10, 11].

There is a discrete series for the Super-Virasoro (or Neveu-Schwarz and Ramond) algebras[1,2,18] and analogs, for the Super-Virasoro, of the character formulas[6,7]. The latter were also used in the proof of the unitarity of the discrete series of the Super-Virasoro[6,7] and in the study of super-conformally invariant theories[19]. I shall also derive formulas for the characters of the super version of the minimal series. In particular, I shall give the proof of the super version of the discrete series characters.

The paper is organized as follows: Sections 2 - 6 deal in detail with the Virasoro algebra. In section 2 the basic definitions and the Kac determinant formula are reviewed. Sections 3 - 5 contain the structural results needed to derive the irreducible characters for general (h,c). This is an alternative approach to the one of [10,11]. The method is a refinement of the one introduced by Wallach and the author in [20,21] where it was successfully applied to derive the irreducible characters in the cases where c = 0, 25 or 26. The main results are the existence of an analytic vector giving embeddings of Verma modules (Theorem 4.4), and a character sum formula for the quotient of two Verma modules (Theorem 5.1). The latter is equivalent to the calculation of the determinant of the hermitian invariant form on a quotient of two Verma modules. Using this formula one derives the irreducible characters by induction on the level of the weight-spaces. The idea of using induction on the level was inspired by the sketch of the proof of the subrepresentation structure theorem of [11]. The character derivation is fully carried out here in the case of the minimal theories and, in particular, in the case of the unitary discrete series representations. Actually, all characters "linked" to these representations are simultaneously determined (Theorem 6.1 and Corollary 6.2). The case of the Super-Virasoro algebras is discussed in §7.

During the present workshop and after the results of this paper were obtained, I became aware of a preprint[22] which contains details of the results of [10,11][1]. The method of proof of the subrepresentation structure theorem is not unrelated to the method presented here, for it relies on the construction of two Jantzen filtrations, an idea first introduced in [20]. Nevertheless, I find that the present approach offers the advantage of leading directly to the character formulas. In particular, the present derivation of the unitary series characters is more transparent than the one that I gave in [8] which requires going through a classification. Finally, I hope that the reader will find it helpful to become acquainted with the present approach in the Virasoro case before proceeding to the super case, where the discussion is more delicate and the notation is more cumbersome.

[1] I thank J.B. Zuber and A. Fialowski for bringing this work to my attention.

The structural results for general h,c, as well as the derivation of the characters of the F-Q-S series over the Virasoro algebra, were presented at the Workshop. In the present written version of the talk I included the derivation of the characters of the B-P-Z series and the case of the Super-Virasoro algebras.

The full details of the general results in the super case will appear in a forthcoming paper[23]. The signature characters of the representations occurring in the minimal theories of Belavin, Polyakov and Zamolodchikov[5] are studied in [24].

ACKNOWLEDGEMENTS

I wish to thank D. Friedan for stimulating conversations. I also wish to thank N. Wallach for an important suggestion.

2. PRELIMINARIES

In this section I shall review the Kac determinant and the definition of irreducible character.

The Virasoro algebra $\underline{g}$ is the complex Lie algebra with basis $\{L_i \mid i \in \mathbb{Z}\} \cup \{L_0'\}$ and bracket relations

$$[L_i, L_j] = (i-j)L_{i+j} + \delta_{i,-j} \frac{i^3-i}{12} L_0' \qquad (1)$$

$$[L_0', \underline{g}] = (0)$$

Let $h,c \in \mathbb{C}$. The Verma module associated with the pair (h,c) is the representation $M = M((h,c))$ of $\underline{g}$ generated by a vector $v = v_{h,c}$, called the vacuum, with precisely the relations:

$$L_i v = 0, \quad i > 0$$

$$L_0 v = hv \qquad (2)$$

$$L_0' v = cv$$

Let M^n be the linear span of the set

$$\{L_{-i_1} L_{-i_2} \cdots L_{-i_k} v \mid i_1 \geq \cdots \geq 0, \ \sum_{j=1}^{k} i_j = n\}.$$

Then $M = \bigoplus_{n \in \mathbb{Z}, n \geq 0} M^n$. Clearly, M^n is the $(h+n)$-eigenspace of L_0 in M and $M^0 = \mathbb{C}v$

There is a symmetric form $(,) = (,)_{h,c}$ on M uniquely defined by:

$$\begin{aligned} &(v,v) = 1 \\ &(L_i w_1, w_2) = (w_1, L_{-i} w_2), \quad i \in \mathbb{Z}, \; w_1, w_2 \in M \\ &(L_0' w_1, w_2) = (w_1, L_0' w_2) \end{aligned} \tag{3}$$

It is easy to see that $(M^n, M^k) = (0)$ if $n \neq k$. Let $(,)_n = (,)_{h,c,n}$ denote the restriction of $(,)$ to M^n. We now state Kac's formula for $\det(,)_n$ [25,26,27] (see also [28]), valid up to a non-zero scalar:

$$\det(,)_n = \prod_{\substack{rs \leq n \\ rs \in \mathbb{N}}} (h - h_{r,s}(c))^{P(n-rs)} \tag{4}$$

where,

$$h_{r,s}(c) = \frac{1}{48}[(13-c)(r^2+s^2) + \sqrt{c^2-26c+25}\,(r^2-s^2) - 24rs - 2 + 2c] \tag{5}$$

and $P(k)$ is the number of ways of writing k as a sum of positive integers.

Let $\mathrm{Rad}(,)$ denote the <u>radical</u> of $(,)$:

$$\begin{aligned} &\mathrm{Rad}(,) = \{w \in M \mid (w,M) = (0)\} \quad \text{and set} \\ &L = L((h,c)) = M/\mathrm{Rad}(,) \end{aligned} \tag{6}$$

It follows from the definition of M and from (3) that L is irreducible.

For a representation W of $\underline{g}$ with an eigenspace decomposition relative to L_0:
$W = \bigoplus_{a \in \mathbb{C}} W^a$, where $W^a = \{w \in W \mid L_0 w = aw\}$, such that $\dim W^a < \infty$ for for all a, we define the (formal) <u>character</u> of W as:

$$\mathrm{ch}\, W = \sum_a (\dim W^a) q^a \tag{7}$$

Clearly,

$$\text{Ch } M = \sum_{n \geq 0} P(n) q^{h+n} = \eta(q)^{-1} q^h \tag{8}$$

where $\eta(q) = \prod_{i=1}^{\infty} (1-q^i)$. The characters ch L are the <u>irreducible characters</u>.

3. CHARACTER SUM FORMULAS FOR VERMA MODULES

The subrepresentation structure of a Verma module over the Virasoro algebra can be obtained from a character sum formula for the quotient of two Verma modules (second character sum formula) (§5). For example, in §6, I will derive the characters of the B-P-Z and F-Q-S series from this formula. In this section, a character sum formula for a Verma module is obtained (first character sum formula). This is an additive version of the determinant formula 2(4). The idea is to consider the family of representations $M((h+z,c))$, $z \in \mathbb{C}$, $z \sim 0$, and view $(\, , \,)_{h+z,c,n}$ as a holomorphic function near (h,c), so that one can study the subrepresentations $M_{(k)}$ of M where $(\, , \,)_{h+z,c,n}$ vanishes with order at least k, for all n. More precisely, following [20] (see also [8]), let $V = \bigoplus_{n \in \mathbb{Z}, n \geq 0} V^n$ where $V^0 = \mathbb{C}$ and, for $n > 0$,

$$V^n = \text{Lin}\{L_{-i_1} \cdots L_{-i_k} \mid i_1 \geq \cdots \geq i_k > 0 \mid \sum_{j=1}^{k} i_j = n\}.$$

(Here, the product of elements of $\underline{g}$ takes place in the universal enveloping algebra $U(\underline{g}) = T(\underline{g})/ \langle X \otimes Y - Y \otimes X - [X,Y] \rangle$). Then, pull back the actions of the spaces $M((h+z,c))$ to the space V via the linear isomorphism $A_z : V \to M((h+z,c))$ defined by $A_z(X) = Xv_{h+z,c}$ for all $X \in V$. That is, set $\pi_z(Y)(X) = A_z^{-1}(Y(A_z X))$ for all $Y \in \underline{g}$, $X \in V$. Then $M((h+z,c)) \simeq (V,\pi_z)$. Let $B_z(w_1,w_2) = (A_z w_1, A_z w_2)_{h+z,c}$ and let $B_{z,n}$ denote the restriction of B_z to V^n. Let $\mathcal{O}(V)$ (respectively. $\mathcal{O}(\mathbb{C})$) denote the space of germs $\underset{\sim}{f}$ of holomorphic functions f at 0 with values in $\bigoplus_{n \in F} V^n$ ($F \subset \mathbb{N} \cup \{0\}$, F finite) (respectively, in $\mathbb{C}$).

For any $k \in \mathbb{Z}$, $k \geq 0$, set

$$\mathcal{O}_{(k)}(V) = \{\underset{\sim}{f} \in \mathcal{O}(V) \mid B_z(f(z),w) \in z^k\mathcal{O}(\mathbb{C}) \text{ for all } w \in V\} \tag{9}$$

Put

$$V_{(k)} = \{f(0) \mid \underset{\sim}{f} \in \mathcal{O}_{(k)}(V)\} \tag{10}$$

and

$$M_{(k)} = A_0(V_{(k)}) \tag{11}$$

Then $M = M_{(0)} \supset M_{(1)} \supset M_{(2)} \supset \ldots$ is a chain of subrepresentations of M, called a Jantzen filtration[9,20]. Up to a non-zero constant, $\det B_{z,n} = a_1(z)\ldots a_r(z)$ with $0 \leq d_1 \leq \ldots \leq d_r < \infty$, where $d_i = \mathrm{ord}_0 a_i(z)$, $1 \leq i \leq r$. One has: $\sum_{k>0} \dim M^n \cap M_{(k)} =$

$$\sum_{k>0} \dim V^n \cap V_{(k)} = \sum_{k>0} \{j \mid k \leq d_j\} = \sum_{j=1}^{r} d_j = \mathrm{ord}_0 \det(\ ,\ )_{h+z,c,n}.$$

This observation implies the first character sum formula:

$$\sum_{k>0} \mathrm{ch}\, M_{(k)} = \sum_{n\in\mathbb{Z},n\geq 0} q^{h+n}\, \mathrm{ord}_0 \det(\ ,\ )_{h+z,c,n} \tag{12}$$
[20]

Let $\phi_i(h,c) = \prod_{rs=i} (h-h_{r,s}(c))$ and set $\mathbf{N}_+((h,c)) = \{i \in \mathbf{N} \mid \phi_i(h,c) = 0\}$, $d(i) = \#\{(r,s) \in \mathbf{N}^2 \mid rs = i\}$. Clearly, $\mathrm{ord}_0(h+z-h_{r,s}(c)) = 1$ or 0 according to whether $h = h_{r,s}(c)$ or $h \neq h_{r,s}(c)$. From (12) and (4) one obtains (13) below.

Proposition 3.1[8]: For any $h,c \in \mathbb{C}$ one has

$$\sum_{k\geq 1} \mathrm{ch}\, M((h,c))_{(k)} = \sum_{i\in \mathbf{N}_+((h,c))} d(i)\,\mathrm{ch}\, M((h+i,c)) \tag{13}$$

Furthermore, $M((h,c))_{(1)} = \mathrm{Rad}(\ ,\ )_{h,c}$

We write $(h',c') \leftarrow (h,c)$ if $c = c'$ and if either $h = h'$ or $h' = h + i$, $\phi_i(h,c) = 0$, for some $i \in \mathbf{N}$. We write $(h',c') \uparrow (h,c)$

if there are (h_i,c_i), $i = 1, \ldots, r$ such that $(h_1,c_1) = (h',c')$, $(h_r,c_r) = (h,c)$ and $(h_i,c_i) \leftarrow (h_{i+1},c_{i+1})$, $i = 1, \ldots, r-1$.

We say that a representation W is a <u>subquotient</u> of a representation V if W = U/U' where $V \supset U \supset U'$ (inclusion of representations).

<u>Corollary 3.2</u>[8]: <u>Let</u> $(h',c'),(h,c) \in \mathbb{C}^2$. <u>If</u> $L((h',c'))$ <u>is a subquotient of</u> $M((h,c))$ <u>then</u> $(h',c') \uparrow (h,c)$.

<u>Notes</u>: 1) Proposition 3.1 is a corrected version of [8, Corollary 2, §2].

2) The meaning of $(h',c') \leftarrow (h,c)$ in the present paper is different from that of the same symbol in [8, §5] and in [10,11].

3) Corollary 3.2 was first stated in [25,26].

4. THE ANALYTIC SHAPOVALOV ELEMENT

The proof of the second character sum formula (§5) relies on the existence of an analytic vector on the variety of the irreducible factors of $\det(\ ,\ )_n$ (Theorem 4.2). The existence theorem is one of the deepest results of this paper. The proof obtained here follows the general lines of argument of [20].

Let

$$\begin{aligned} \psi_{r,s}(h,c) &= (h-h_{r,s}(c))(h-h_{s,r}(c)), \quad r \neq s \\ \psi_{r,r}(h,c) &= h + \frac{1}{24}(r^2-1)(c-1) \end{aligned} \tag{14}$$

$\psi_{r,s}(h,c)$ is an irreducible polynomial of second degree (respectively, of first degree) in h and c for $r \neq s$ (respectively, for $r = s$). Define the varieties in $\mathbb{C}^2$.

$$V_{r,s} = \{(h,c) \in \mathbb{C}^2 \mid \psi_{r,s}(h,c) = 0\} \tag{15}$$

Choose the parameter $t = \dfrac{13 - c + \sqrt{c^2-26c+25}}{12}$ and set

$$\begin{aligned} h_{r,s}(t) &= \frac{(rt-s)^2 - (t-1)^2}{4t} \\ c(t) &= 1 - \frac{6(t-1)^2}{t} \end{aligned} \tag{16}$$

It is immediate from (5), (14) and (15) that

$$V_{r,s} = \{(h_{r,s}(t), c(t)) \mid t \in \mathbb{C} \setminus \{0\}\} \tag{17}$$

Note: One can also use the parametrization given in [1,2]. The parametrization above, which was also used in [11,28], turns out to be more convenient for the derivation of the B-P-Z characters.

Let

$$F_{r,s} = V_{r,s} \cap \Big(\bigcup_{\substack{r's' \leq rs \\ (r',s') \neq (r,s)}} V_{r',s'} \Big) \tag{18}$$

Using the rational parametrization of $V_{r,s}$ given by (16) one obtains:

Proposition 4.1: $F_{r,s}$ is finite.

Let

$$V'_{r,s} = V_{r,s} \setminus F_{r,s} \tag{19}$$

Theorem 4.2: (i) For all $(h,c) \in V_{r,s}$,

$$\dim \mathrm{Rad}(\ ,\ )_{h,c,rs} \leq \sum_{i \in N_+((h,c))} d(i) P(rs-i);$$

(ii) for all $(h,c) \in V'_{r,s}$, $\dim \mathrm{Rad}(\ ,\ )_{h,c,rs} \leq 1$;

(iii) if rs is minimum with the property that $\det(\ ,\ )_{h,c,rs} = 0$ then the equality holds in (ii).

Proof: One has, for $r \neq s$,

$$\frac{\partial}{\partial h} \psi_{r,s}(h,c)(h_{r,s}(t), c(t)) = \frac{(s^2-r^2)(t^2-1)}{4t}$$

and

$$\frac{\partial}{\partial c} \psi_{r,s}(h,c)(h_{r,s}(t)(c(t)) = \frac{[(1-r^2)t^2-(1-s^2)](s^2-r^2)}{24t}$$

Hence,

$$\mathrm{grad}\ \psi_{r,s}(h,c)(h_{r,s}(t), c(t)) \neq (0,0)$$

for all t.

The result now follows from Proposition 3.1 and by the local diagonalization argument of the proof of (12) (see observation preceeding (12)) applied to the first variable when

$$\frac{\partial}{\partial h}\psi_{r,s}(h,c)(h_{r,s}(t),c(t)) \neq 0$$

and to the second variable, otherwise. Q.E.D.

Remark: The implication of the non-vanishing of the gradient was suggested to the author by Wallach.

Scholium 4.3: Let $(h_0,c_0) \in V_{r,s}$. Set $u = \psi_{r,s}(h,c), v = c$, if $\frac{\partial}{\partial h}\psi_{r,s}(h,c)(h_0,c_0) \neq 0$. Otherwise, choose $u = h$, $v = \psi_{r,s}(h,c)$. Then $\{u,v\}$ is a system of coordinates near (h_0,c_0).

The following is the extension of [20, Theorem A] to arbitrary values of the central charge.

Theorem 4.4: Let $(h,c) = (h_{r,s}(t_0), c(t_0)) \in V_{r,s}$, where rs is minimum with the property that $\det(\ ,\ )_{h,c,rs} = 0$. There is $\varepsilon > 0$ and a holomorphic map $\xi^{h,c}_{r,s}: D_\varepsilon((h,c)) \to V^{rs}$ such that for t near t_0 $\xi^{h,c}_{r,s}(h_{r,s}(t),c(t))v$ gives the embedding

$$M((h_{r,s}(t) + rs, c(t)) \hookrightarrow M((h_{r,s}(t),c(t))$$

(Here $D_\varepsilon((h,c)) = \{(h'c') \in \mathbb{C}^2 \,|\, |h'-h| < \varepsilon, |c'-c| < \varepsilon\}$.)

Proof: Let (h,c) satisfy the hypothesis conditions. Proceeding as in [20, Theorem A] the result holds for all (h,c) not in $F_{r,s}$ by Theorem 4.2(iii) and the Scholium. By Proposition 4.1 the result follows for all (h,c). Q.E.D.

5. CHARACTER SUM FORMULAS FOR QUOTIENTS OF VERMA MODULES

I will now derive the second character sum formula. The formula says that the determinant of the form on a quotient of Verma modules is the quotient of the determinants of the forms on each Verma module. This formula gives the induction step for proving character formulas (see §6).

We now assume that $\det(\ ,\ )_{h,c,n} = 0$ for some n, otherwise chL = chM which is given in §2. Let $(h,c) = (h_{r,s}(t_0), c(t_0)) \in V_{r,s}$ where rs is minimum with the property that $\det(\ ,\ )_{h,c,rs} = 0$.

Set $N = N((h,c)) = M((h,c))/M((h+rs,c))$ and let $(\ ,\ )_N = (\ ,\ )_{N,h,c}$ denote the form induced by $(\ ,\ )_{h,c}$ on N.

Theorem 5.1: There is a chain of subrepresentations

$$N = N_{(0)} \supset N_{(1)} \supset \ldots \text{ such that}$$

(i) $N_{(1)} = \mathrm{Rad}(\ ,\ )_N$

(ii) $$\sum_{i>0} \mathrm{ch} N_{(i)} = \sum_{i \in \mathbb{N}_+((h,c))} d(i)\mathrm{ch}M((h+i,c)) - d(rs)\mathrm{ch}M((h+rs,c)) - \sum_{j \in \mathbb{N}_+((h+rs,c))} d(j)\mathrm{ch}M((h+rs+j,c)) \tag{20}$$

Proof: Let $S_{t_0} = \{t \mid (h_{r,s}(t), c(t)) \in D_\varepsilon((h,c))$, with $D_\varepsilon((h,c))$ as in §4, Theorem 2. Set

$$N((h_{r,s}(t),c(t)) = M((h_{r,s}(t),c(t))/\xi^{h,c}_{r,s}(h_{r,s}(t),c(t))v.$$

There is $U \subset V$ such that

$$V = U \oplus V\, \xi^{h,c}_{r,s}(h_{r,s}(t),c(t)) \quad \text{for all } t \in S_{t_0}.$$

Using U in place of V and $(\ ,\ )_N$ in place of $(\ ,\ )$ one obtains a chain of subrepresentations $N = N_{(1)} \supset N_{(2)} \supset \ldots$ with the property (i) above and such that

$$\sum_{i>0} \mathrm{ch} N_{(i)} = \sum_{n \in \mathbb{N}, n \geq 0} q^{h+n} \mathrm{ord}_{t_0} \det(\ ,\ )_{N,h_{r,s}(t),c(t),n} \tag{21}$$

Proceeding as in [20, §4] but using Theorems 4.2 and 4.4, and (21) in place of [20, Theorems A and B] one obtains

$$\sum_{i>0} \mathrm{ch} N_{(i)} = \sum_{i>0} \mathrm{ch}M((h,c))_{(i)} - d(rs)\mathrm{ch}M((h+rs,c)) - \sum_{j>0} \mathrm{ch}M((h+rs,c))_{(j)} \tag{22}$$

(20) now follows from (22) and (13). Q.E.D.

Note: The Jantzen filtration for quotients, $\{N_{(i)}\}$, in the case $c = 0$ was first constructed in [20].

6. THE CHARACTERS OF THE MINIMAL AND UNITARY SERIES

Consider the series of representations $L((h,c))$ where

$$c = c((p,p')) = 1 - 6\,\frac{(p-p')^2}{pp'},\quad p,p' \in \mathbb{N},\ \text{g.c.d.}(p,p') = 1$$
$$h = h_{r_0,s_0}((p,p')) = \frac{(pr_0-p's_0)^2-(p-p')^2}{4pp'} \tag{23}$$

$$(1 \le r_0 \le p'-1,\ 1 \le s_0 \le p-1,\ s_0p' < r_0p)$$

Note that $(h_{r,s}((p,p')),\ c((p,p'))) = h_{r,s}(\frac{p}{p'}),\ c(\frac{p}{p'}))$ where $(h_{r,s}(t),c(t))$ is defined by (16). This series was proposed in [5] to define the minimal conformal theories in statistical mechanics (see [17]). Note that the unitary series of [1,2] is given by (23) as the special case where $p = m+1$, $p' = m$.

For all $k \in \mathbb{Z}$, set

$$a_k = a_k(p,p',r_0,s_0) = \frac{(pr_0+p's_0+2pp'k)^2 - (p-p')^2}{4pp'}$$
$$b_k = b_k(p,p',r_0,s_0) = \frac{(pr_0-p's_0+2pp'k)^2 - (p-p')^2}{4pp'} \tag{24}$$

Fix $c = c((p,p'))$. For simplicity I will, from now on, drop the second variable in the notations $M((h,c))$, $L((h,c))$ and $N((h,c))$.

Theorem 6.1: Let $k \in \mathbb{Z}$, $k \ge 0$. One has

$$\text{ch}L(b_{\pm k}) = \eta(q)^{-1}\Big(q^{b_{\pm k}}-q^{a_k}+\sum_{\substack{\ell\in\mathbb{Z}\\ |\ell|>k}}(-q^{a_\ell}+q^{b_\ell})\Big) \tag{25}$$

$$\text{ch}L(a_k) = \eta(q)^{-1}\Big(q^{a_k}+q^{a_{k+1}}+\sum_{\substack{\ell,\ell'\in\mathbb{Z}\\ |\ell|>k,\,|\ell'|>k+1}}(-q^{b_\ell}+q^{a_{\ell'}})\Big) \tag{26}$$

$$\text{ch}L(a_{-(k+1)}) = \eta(q)^{-1}\Big(q^{a_{-(k+1)}}+q^{a_{k+1}}+ + \sum_{\substack{\ell,\ell'\in\mathbb{Z}\\ |\ell|>k,\,|\ell'|>k+1}}(-q^{b_\ell}+q^{a_{\ell'}})\Big) \tag{27}$$

Corollary 6.2: Let h be as in (23). Then

$$\operatorname{ch}L(h) = \eta(q)^{-1} \sum_{k \in \mathbb{Z}} (q^{b_k} - q^{a_k}) \tag{28}$$

Remark: For $p = m+1$ and $p' = m$, $m \in \mathbb{Z}$, $m \geq 2$, (28) gives the unitary discrete series characters[8].

Proof of Theorem 6.1: I will begin by calculating the right-hand side of (13) for $h \in \{b_{\pm k}, a_{\pm k}, k \in \mathbb{Z}, k \geq 0\}$. It is more convenient to use the following version of the determinant formula:

$$\det(\ ,\ )_{h,c,n} = \prod_{rs \leq n} (h - h_{r,s}(\tfrac{p}{p'}))^{P(n-rs)} \tag{29}$$

where $h_{r,s}(t)$ is given by (16). Let $\phi_i((h,c)) = 0$. Then there are $r,s \in \mathbb{N}$, with $rs = i$, such that, either

$$2pp'k + pr_0 - p's_0 = pr - p's$$

or

$$2pp'k + pr_0 - p's_0 = p's - pr$$

This implies that, either

$$\left.\begin{aligned} r &= r_0 + 2p'k + p'\ell \\ s &= s_0 + p\ell \end{aligned}\right\} , \ \ell \geq 0,\ \ell \in \mathbb{Z} \tag{30}$$

or

$$\left.\begin{aligned} r &= -r_0 - 2p'k + p'\ell \\ s &= -s_0 + p\ell \end{aligned}\right\} , \ \ell \in \mathbb{Z},\ \ell > 0 \tag{31}$$

(30) and (31) give, respectively

$$b_k + i = a_{(\ell+k)} \quad , \ \ell \in \mathbb{Z},\ \ell \geq 0 \tag{32}$$

and

$$b_k + i = a_{-(\ell+k)} \quad , \ \ell \in \mathbb{Z},\ \ell > 0 \tag{33}$$

Similarly, for $h = b_{-k}$, $h = a_{\pm k}$. Also note that $a(i) = 1$ if $i \in \mathbb{N}_+((h,c))$, and $h = a_{\pm k}$ or $h = b_{\pm k}$. Summarizing, (13) gives

$$\sum_{i>0} \operatorname{ch}M(b_{\pm k})(i) = \sum_{\ell \in \mathbb{Z}, \ell \geq 0} \operatorname{ch}M(a_{k+\ell}) + \operatorname{ch}M(a_{-(k+\ell+1)}) \tag{34}$$

$$\sum_{i>0} \mathrm{chM}(a_k)_{(i)} = \sum_{i>0} \mathrm{chM}(a_{-(k+1)})_{(i)} = \sum_{\ell \in \mathbb{N}} (\mathrm{chM}(b_{k+\ell}) + \mathrm{chM}(b_{-(k+\ell)})) \tag{35}$$

It is not hard to see that $N(b_{\pm k}) = M(b_{\pm k})/M(a_k)$, $N(a_k) = M(a_k)/M(b_{-(k+1)})$ and $N(a_{-k}) = M(a_{-k})/M(b_{-k})$.

The description of N can be visualized by the following partial diagram of the relation $(h,c) \leftarrow (h',c')$ (abbreviated to $h \leftarrow h'$) defined in §3. The diagram shows each weight (eigenvalue of L_0) connected to the closest linked weight.

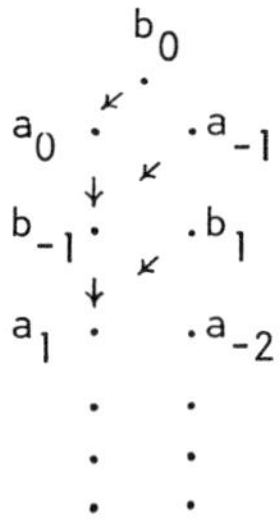

$$\begin{array}{ccc} & & b_0 \\ & \swarrow & \\ a_0 & & a_{-1} \\ \downarrow & \swarrow & \\ b_{-1} & & b_1 \\ \downarrow & \swarrow & \\ a_1 & & a_{-2} \\ \vdots & & \vdots \end{array}$$

Substituting (34) and (35) in (22) yields:

$$\sum_{i>0} \mathrm{chN}(b_{\pm k})_{(i)} = \sum_{\ell \in \mathbb{Z}, \ell \geq 0} (\mathrm{chM}(a_{k+\ell}) + \mathrm{chM}(a_{-(k+\ell+1)})) - \mathrm{chM}(a_k) - \sum_{\ell \in \mathbb{N}} (\mathrm{chM}(b_{k+\ell}) + \mathrm{chM}(b_{-(k+\ell)})) \tag{36}$$

$$\sum_{i>0} \mathrm{chN}(a_k)_{(i)} = \sum_{\ell \in \mathbb{N}} (\mathrm{chM}(b_{k+\ell}) + \mathrm{chM}(b_{-(k+\ell)})) - \mathrm{chM}(b_{-(k+1)}) - \sum_{\ell \in \mathbb{N}} (\mathrm{chM}(a_{k+\ell}) + \mathrm{chM}(a_{-(k+\ell+1)})) \tag{37}$$

$$\sum_{i>0} \mathrm{chN}(a_{-k})_{(i)} = \sum_{\ell \in \mathbb{Z}, \ell \geq 0} (\mathrm{chM}(b_{k+\ell}) + \mathrm{chM}(b_{-(k+\ell)})) - \mathrm{chM}(b_{-k}) - \sum_{\ell \in \mathbb{Z}, \ell \geq 0} (\mathrm{chM}(a_{k+\ell}) + \mathrm{chM}(a_{-(k+\ell+1)})) \tag{38}$$

The next step is to prove the following formulas:

$$\begin{aligned} \operatorname{ch} M(b_{\pm k})_{(1)} &= \operatorname{ch}(M(a_k)+M(a_{-(k+1)})) = \operatorname{ch}(M(b_k)\cap M(b_{-k})) \\ \operatorname{ch} M(a_k)_{(1)} &= \operatorname{ch} M(a_{-(k+1)})_{(1)} = \operatorname{ch}(M(b_{-(k+1)})+M(b_{k+1})) = \\ &\operatorname{ch}(M(a_k)\cap M(a_{-(k+1)})) \end{aligned} \tag{39}$$

or, equivalently, for all $n \in \mathbb{Z}$, $n \geq 0$,

$$\begin{aligned} \dim M(b_{\pm k})_{(1)}^{b_{\pm k}+n} &= \dim(M(a_k)+M(a_{-(k+1)}))^{b_{\pm k}+n} = \\ &= \dim(M(b_k)\cap M(b_{-k}))^{b_{\pm k}+n} \\ \dim M(a_k)_{(1)}^{a_k+n} &= \dim(M(b_{-(k+1)})+M(b_{k+1}))^{a_k+n} \\ &= \dim(M(a_k)\cap M(a_{-(k+1)}))^{a_k+n} \\ \dim M(a_{-(k+1)})^{a_{-(k+1)}+n} &= \dim(M(b_{-(k+1)})+M(b_{k+1}))^{a_{-(k+1)}+n} \\ &= \dim(M(a_k)\cap M(a_{-(k+1)}))^{a_{-(k+1)}+n} \end{aligned} \tag{A(n)}$$

Clearly, the statement (A(0)) is valid. Proceeding by induction on n, assume that (A(m)) is valid for all $m < n$. Now write $b_{\pm k}+n = a_k+(n+b_{\pm k}-a_k)$, $a_k+n = b_{\pm(k+1)}+(n+a_k-b_{\pm(k+1)})$, $a_{-(k+1)} = b_{\pm(k+1)}+(n+a_{-(k+1)}-b_{\pm(k+1)})$, and note that $b_{\pm k} - a_k$, $a_k - b_{\pm(k+1)}$ and $a_{-(k+1)} - b_{\pm(k+1)}$ are all negative integers. From (36) - (38) and the induction hypothesis one obtains:

$$\sum_{i>0} \dim N(b_{\pm k})_{(i)}^{b_k+n} = \dim(M(a_k)+M(a_{-(k+1)}))^{b_k+n} - \dim M(a_k)^{b_k+n} \tag{40}$$

$$\sum_{i>0} \dim N(a_k)_{(i)}^{a_k+n} = \dim(M(b_{k+1})+M(b_{-(k+1)}))^{a_k+n} - \dim M(b_{-(k+1)})^{a_k+n} \tag{41}$$

$$\sum_{i>0} \dim N(a_{-(k+1)})_{(i)}^{a_{-(k+1)}+n} = \dim(M(b_{k+1})+M(b_{-(k+1)}))^{a_{-(k+1)}+n} - \dim M(b_{-(k+1)})^{a_{-(k+1)}+n} \tag{42}$$

Now, the first term of the sum on the left-hand side of each of the above equations (40) - (42) is at least equal to the right-hand side. Hence,

$$\dim N(b_{\pm k})_{(1)}^{b_k+n} = \dim(M(a_k)+M(a_{k+1}))^{b_{\pm k}+n} - \dim M(a_k)^{b_{\pm k}+n} \tag{43}$$

$$\dim N(a_k)_{(1)}^{a_k+n} = \dim(M(b_{-(k\underline{+}1)})+M(b_{k+1}))^{a_k+n} - \dim M(b_{-(k+1)})^{a_k+n} \tag{44}$$

$$\dim N(a_{-(k+1)})_{(1)}^{a_{-(k+1)}+n} = \dim(M(b_{-(k+1)})+M(b_{k+1}))^{a_{-(k+1)}+n} - \dim M(b_{-(k+1)})^{a_{-(k+1)}+n} \tag{45}$$

Therefore, (A(n)) is valid for all n, i.e., (39) holds. The theorem follows from (39).

7. CHARACTERS IN THE SUPER CASE

We now extend the results of sections 3 - 6 to the case of the super-Virasoro algebras. The notation used here is independent of §§2 - 6. The details of the structural theorems will appear in [23].

The super-Virasoro algebra $\tilde{\underline{g}}$ is defined by the generators $\{L_0', L_i, G_j\}$ and relations[29,30]

$$\begin{aligned} [L_i, L_j] &= (i-j)L_{i+j} + \delta_{i,-j}\left(\frac{i^3-i}{8}\right)L_0' \\ [G_i, G_j] &= 2L_{i+j} + \delta_{i,-j}\frac{1}{2}\left(i^2-\frac{1}{4}\right)L_0' \\ [L_i, G_j] &= \left(\frac{i}{2}-j\right)G_{i+j} \\ [L_0', \tilde{\underline{g}}] &= (0) \end{aligned} \tag{46}$$

The L_i are indexed by $\mathbb{Z}$ as in section 2, while the G_i are indexed alternatively by $\frac{1}{2}+\mathbb{Z}$ or by $\mathbb{Z}$. In the first case $\tilde{\underline{g}} = \tilde{ns}$ is called the

<u>Neveu-Schwarz algebra</u> and in the second case $\hat{\underline{g}} = \hat{\underline{r}}$ is called the <u>Ramond algebra</u>.

Let $h, c \in \mathbb{C}$, $\hat{c} = \frac{2}{3}c$. The <u>Verma module</u> associated with the pair $(h, \hat{c})$ is defined as in §2, as the $\mathbb{Z}_2$-graded representation $M = M((h,\hat{c})) = M_{\bar{0}} \oplus M_{\bar{1}}$ of $\tilde{\underline{g}}$ generated by a vector $v = v_{h,\hat{c}}$ of degree 0, called the vacuum, with exactly the relations:

$$\begin{aligned} L_i v &= G_i v = 0, \quad i > 0 \\ L_0 v &= hv \\ L_0' v &= \hat{c} v \end{aligned} \tag{47}$$

Here the degree is given by the <u>chirality operator</u> $\Gamma v = v$, $[\Gamma, G_n]_+ = [\Gamma, L_n] = 0$[18], where $[A,B]_+$ denotes the anticommutator $AB + BA$, via the correspondence: $M_{\bar{0}}$ = 1-eigenspace of Γ, $M_{\bar{1}}$ = -1-eigenspace of Γ.

There is a symmetric form $(\, , \,) = (\, , \,)_{h,c}$ on M defined similarly to the one in §2 with the additional provision that $(\Gamma w_1, w_2) = (w_1, \Gamma w_2)$ for all $w_1, w_2 \in M$.

One has the decomposition $M = \bigoplus_{n \in Z, n \geq 0} M^n$, as before, where M^n is the linear span of $\{L_{-i_1} \dots L_{-i_k} G_{-j_1} \dots G_{-j_\ell} v \mid i_1 \geq \dots \geq i_k \geq 0, j_1 > \dots > j_\ell \geq 0, \sum_{t=1}^{k} i_t + \sum_{u=1}^{\ell} j_u = n\}$, and $Z = \frac{1}{2}\mathbb{Z}$ if $\hat{\underline{g}} = \underline{\hat{ns}}$, $Z = \mathbb{Z}$, if $\hat{\underline{g}} = \hat{\underline{r}}$. Let $M_\gamma^n = M^n \cap M_\gamma$, $\gamma \in \{\bar{0}, \bar{1}\}$. Denote by $(\, , \,)_{n,\gamma} = (\, , \,)_{h,c,n,\gamma}$ the restriction of $(\, , \,)_n$ to M_γ^n if $M_\gamma^n \neq \phi$

Let $\varepsilon = 0$ if $\hat{\underline{g}} = \underline{\hat{ns}}$ and $\varepsilon = \frac{1}{2}$ if $\hat{\underline{g}} = \hat{\underline{r}}$. Let $r, s \in \mathbb{Z}$, $r, s \geq 0$, with $r - s \in 2\varepsilon + 2\mathbb{Z}$. Up to a nonzero scalar, one has, for $n > 0$

$$\det(\, , \,)_{n,\gamma} = (h - \tfrac{\hat{c}}{16})^{\varepsilon \hat{P}_\varepsilon(n)} \prod_{\frac{rs}{2} \leq n} (h - h_{r,s}(\hat{c}))^{\hat{P}_\varepsilon(n - rs/2)} \tag{48}$$

$$\det(\, , \,)_{0,\bar{0}} = 1, \quad \det(\, , \,)_{0,\bar{1}} = h - \frac{\hat{c}}{16} \tag{49}$$

where

$$h_{r,s}(\hat{c}) = \frac{1}{32}[(r^2+s^2)(5-\hat{c}) + \sqrt{\hat{c}^2-10\hat{c}+9}(r^2-s^2) - 2rs - 2 + 2\hat{c}] + \frac{\varepsilon}{8} \tag{50}$$

and

$$\sum_{i \in (\frac{1}{2}+\varepsilon)\mathbb{Z}_+} \hat{P}_\varepsilon(i)q^i = \prod_{n \in \mathbb{N}} \frac{(1+q^{n+\varepsilon-1/2})}{(1-q^n)} \tag{51}$$

The determinant formulas for $n\hat{s}$ and $\hat{r}$ were conjectured by Kac[26] and Friedan, Qiu and Shenker[18,31], respectively. Their proofs were given by Meurman and the author[32].

The character is defined similarly to the character in the Virasoro case. One has

$$\mathrm{ch}M = (1-\varepsilon)^{-1} q^h \prod_{n \in \mathbb{N}} \frac{(1+q^{n+\varepsilon-1/2})}{(1-q^n)} \tag{52}$$

Following [32] let $\phi_i(h,\hat{c}) = \prod_{\frac{rs}{2}=i} (h-h_{r,s}(\hat{c}))$ where $r,s \in \mathbb{Z}$, $r,s \geq 0$, $r-s \in 2\varepsilon + 2\mathbb{Z}$. Let $(\frac{1}{2}+\varepsilon)\mathbb{N}_+((h,\hat{c})) = \{i \in (\frac{1}{2}+\varepsilon)\mathbb{N} \mid \phi_i(h,\hat{c})=0\}$. Let $\hat{d}(i) = \{(r,s) \in \mathbb{N} \times \mathbb{N} \mid \frac{rs}{2} = i$ where $r-s \in 2\varepsilon + 2\mathbb{Z}\}$.

Proposition 7.1 [32, Proposition 5.3]: There is a filtration $M = M_{(0)} \supset M_{(1)} \supset \ldots$ of Γ-invariant subrepresentations of M such

that $$\mathrm{Rad}(\ ,\ ) = M_{(1)} \tag{53}$$

and $$\sum_{k \geq 1} \mathrm{ch}M((h,\hat{c}))_{(k)} = \varepsilon\delta_{h,\hat{c}/16}\, \mathrm{ch}M(h,\hat{c})) + \sum_{i \in (1/2+\varepsilon)\mathbb{N}_+((h,\hat{c}))} \hat{d}(i)\,\mathrm{ch}M((h+i,\hat{c})) \tag{54}$$

Let $\psi_{r,s}(h,\hat{c}) = (h-h_{r,s}(\hat{c}))(h-h_{s,r}(\hat{c}))$, $r,s \in \mathbb{Z}$, $r,s \geq 0$, $r-s \in 2\varepsilon + 2\mathbb{Z}$, $r \neq s$, $\psi_{r,r}(h,\hat{c}) = h - \frac{1}{16}(1-\hat{c})(r^2-1)$. Let $V_{r,s}(h,\hat{c}) = \{(h,\hat{c}) \in \mathbb{C}^2 \mid \psi_{r,s}(h,\hat{c}) = 0$, $r,s \in \mathbb{Z}$. $r,s \geq 0$, $r-s \in 2\varepsilon + 2\mathbb{Z}$. Then one has[32]

$$V_{r,s} = \{(h_{r,s}(t),\hat{c}(t)) \mid t \in \mathbb{C} \setminus \{0\}\} \tag{55}$$

where

$$\hat{c}(t) = 1 - \frac{2(t-1)^2}{t}$$
$$h_{r,s}(t) = \frac{(rt-s)^2 - (t-1)^2}{8t} + \frac{\varepsilon}{8} \qquad (56)$$

As in §5, one sets $N = N((h,\hat{c})) = M((h,\hat{c}))/M((h+rs,\hat{c}))$ where $rs \neq 0$ and rs is minimum with the property that $\det(\ ,\ )_{h,\hat{c},rs} = 0$. Let $(\ ,\ )_N$ denote the form induced by $(\ ,\ )$ on N. Using the parametrization (56) of the variety of irreducible factors of $\det(\ ,\ )$, instead of (16), one obtains:

Theorem 7.2 [23]: There is a chain of Γ-invariant subrepresentations $N = N_{(0)} \supset N_{(1)} \supset \ldots$ such that

(i) $N_{(1)} = \mathrm{Rad}(\ ,\ )_N$

(ii) $$\sum_{i>0} \mathrm{ch}N_{(i)} = \varepsilon\delta_{h,\frac{\hat{c}}{16}}\mathrm{ch}M((h,\hat{c})) + \sum_{i\in(\frac{1}{2}+\varepsilon)N_+((h,\hat{c}))} \hat{d}(i)\,\mathrm{ch}M((h+i,\hat{c}))$$
$$- \hat{d}(rs)\,\mathrm{ch}M((h+rs,\hat{c})) \qquad (57)$$
$$- \sum_{j\in(\frac{1}{2}+\varepsilon)N_+((h+rs,\hat{c}))} \hat{d}(j)\,\mathrm{ch}M((h+rs+j,\hat{c}))$$

Let $L((h,\hat{c})) = M((h,\hat{c}))/\mathrm{Rad}(\ ,\ )$ as before. $L((h,\hat{c}))$ is irreducible as a Γ-invariant representation.

We now consider the super version of the minimal theory series[2]:

$$\hat{c} = \hat{c}((p,p')) = 1 - 2\frac{(p-p')^2}{pp'}, \quad p,p' \in 2\mathbb{N}-1,\ \text{g.c.d.}\ (p,p') = 1, \text{ or}$$
$$p,p' \in 2\mathbb{N},\ \frac{p-p'}{2} \in 2\mathbb{N}-1,\ \text{g.c.d.}\left(\frac{p}{2},\frac{p'}{2}\right) = 1$$
$$h = h_{r,s}((p,p')) = \frac{(pr-p's)^2 - (p-p')^2}{8pp'} + \frac{\varepsilon}{8},\ 1 \le r \le p'-1,\ 1 \le s \le p-1 \qquad (58)$$
$$r-s \in 2\varepsilon + 2\mathbb{Z}$$

---

[2]See [33] for the case $\hat{\underline{g}} = \underline{\hat{ns}}$.

Note that $(h_{r,s}((p,p')), \hat{c}((p,p'))) = (h_{r,s}(\frac{p}{p'}), \hat{c}(\frac{p}{p'}))$ where $(h_{r,s}(t), \hat{c}(t))$ is given by (56), and that the super version of the F-Q-S discrete series[1,2,18] corresponds to (58) where $p = m+2$, $q = m$, $m = 2,3, \ldots$. For all $k \in \mathbb{Z}$, set

$$\begin{aligned} a_k &= a_k(p,p',r,s) = \frac{(pr+p's+2pp'k)^2-(p-p')^2}{8pp'} + \frac{\varepsilon}{8} \\ b_k &= b_k(p,p',r,s) = \frac{(pr-p's+2pp'k)^2-(p-p')^2}{8pp'} + \frac{\varepsilon}{8} \end{aligned} \tag{59}$$

Fix $\hat{c} = \hat{c}((p,p')$ and write $L(h)$ for $L((h,\hat{c}))$, as before. Proceeding as in §6, but using Theorem 7.2 instead of Theorem 5.1, one obtains formulas analogous to (25) - (27). In particular, one has

$$\mathrm{ch}L(h) = (1-\varepsilon)^{-1}\eta_\varepsilon(q) \sum_{k\in\mathbb{Z}} (q^{b_k}-q^{a_k}) \tag{60}$$

where

$$\eta_\varepsilon(q) = \prod_{n\in\mathbf{N}} \frac{(1+q^{n+\varepsilon-1/2})}{(1-q^n)}$$

In the special case where $p = m+2$, $p' = m$, $m = 2,3, \ldots$, (60) gives the characters of the super unitary series.

REFERENCES

[1] Friedan, D., Qiu, Z. and Shenker, S.H., in: Vertex Operators in Mathematics and Physics, Proceedings of a Conference, November 10 - 17, 1983, ed. J. Lepowsky, S. Mandelstam and I. Singer, M.S.R.I. 3, Springer, Berlin-New York (1984).

[2] Friedan, D., Qiu, Z. and Shenker, S.H., Phys. Rev. Lett. 52, 1575 (1984).

[3] Friedan, D., Qiu, Z. and Shenker, S.H., Details of the Non-unitary Proof for Highest Weight Representations of the Virasoro Algebra, Preprint (1986).

[4] Langlands, R., On Unitary Representations of the Virasoro Algebra, Preprint (1986).

[5] Belavin, A.A., Polyakov, A.M. and Zamolodchikov, A.B., Nucl. Phys. B241, 333 (1984).

[6] Goddard, P., Kent, A. and Olive, D., Commun. Math. Phys. 103, 105 (1986).

[7] Kac, V. and Wakimoto, M., in Proceedings of the Symposium on Conformal Groups and Structures, Clausthal Notes in Physics 261 (1986).

[8] Rocha-Caridi, A., in: Vertex Operators in Mathematics and Physics, Proceedings of a Conference, November 10 - 17, 1983, ed. J. J. Lepowsky, S. Mandelstam and I. Singer, M.S.R.I. 3, 451, Springer, Berlin-New York (1984).

[9] Rocha-Caridi, A. and Wallach, N.R., Transactions of the A.M.S. 277, No. 1, 133(1983).

[10] Feigin, B.L. and Fuchs, D.B., Functs. Anal. Prilozhen. 17, No. 3, 91(1983).

[11] Feigin, B.L. and Fuchs, D.B., Topology (Leningrad, 1982). Lecture Notes in Math. 1060, 230, Springer, Berlin-New York (1984).

[12] Cardy, J., Nucl. Phys. B270, 186(1986).

[13] Cardy, J., Effect of Boundary Conditions on the Operator Content of Two-dimensional Conformally Invariant Theories, Preprint (1986).

[14] Itzykson, C., and Zuber, J.-B., Two-dimensional Conformal Invariant Theories on a Torus, Nucl. Phys., to appear.

[15] Cappelli, A., Itzykson, C. and Zuber, J.-B., Modular Invariant Partition Functions, Nucl. Phys., to appear.

[16] Cardy, J., Conformal Invariance in Critical Systems with Boundaries, these proceedings.

[17] Zuber, J.-B., Conformal Invariant Theories, Modular Invariance and Symmetries, these proceedings.

[18] Friedan, D., Qiu, Z. and Shenker, S.H., Phys. Lett. 151B, 37 (1985).

[19] Kastor, D., Modular Invariance in Superconformal Models, Preprint (1986).

[20] Rocha-Caridi, A. and Wallach, N.R., Invent. math. 72, 57(1983).

[21] Rocha-Caridi, A. and Wallach, N.R., Math. Z 185, 1(1984).

[22] Feigin, B.L., Fuchs, D.B., Representations of the Virasoro Algebra, Preprint.

[23] Rocha-Caridi, A., Verma Modules over the Super-Virasoro Algebras, in preparation.

[24] Rocha-Caridi, A. and Wallach, N.R., Signature Characters of Representations of the Virasoro Algebra, in preparation.

[25] Kac, V., Highest Weight Representations of Infinite Dimensional Lie Algebras, Proceedings of the I.C.M., 299, Helsinki(1978).

[26] Kac, V., Contravariant Form for Infinite Dimensional Lie Algebras and Superalgebras, Lecture Notes in Physics, 94, 441(1979).

[27] Feigin, B.L. and Fuchs, D.B., Functs. Anal. Prilozhen. 16, No. 2, 47(1982).

[28] Thorn, C.B., Nucl. Phys. B248, 551(1984).

[29] Neveu, P., Schwarz, J.H., Nucl. Phys. B31, 86(1971)

[30] Ramond, P., Phys. Rev. D3, 2415(1971)

[31] Friedan, D., Letter to author (1984).

[32] Meurman, A. and Rocha-Caridi, A., Highest Weight Representations of the Neveu-Schwarz and Ramond Algebras, Commun. in Math. Phys., to appear.

[33] Bershadsky, M.A., Knizhnik, V.G. and Teitelman, M.G., Phys. Lett. 151B, No. 1, 31(1985).

CONFORMAL INVARIANCE IN CRITICAL SYSTEMS WITH BOUNDARIES

John L. Cardy

Department of Physics, University of California
Santa Barbara, CA 93106
USA

ABSTRACT

The consequences of conformal invariance for systems with free or fixed boundary conditions are presented. It is shown that the eigenvalues of the Virasoro generator L_0 in the semi-infinite geometry are related to the surface critical exponents. For unitary models with $c < 1$ the complete set of such exponents can be derived using a generalization of modular invariance.

Any real physical sample has boundaries. Close to a critical point these boundaries give rise to a significant distortion of the correlation functions, which implies that results of a scattering experiment are related in a rather complicated way to bulk critical properties. In two dimensions, samples tend to be rather small, and these effects are expected to be important. Conformal invariance is a useful tool in understanding them. At an elementary level, one may use conformal mappings to calculate correlation functions in some arbitrary region from knowledge of their behavior in the half-plane. Such calculations have been performed for the Ising model[1]. However, there is a deeper application of conformal invariance. Just as this principle, when augmented by extra physical requirements like positivity[2] and modular invariance[3], gives strong restrictions on possible bulk universality classes, so also does it have important consequences for the behavior near a boundary[4,5]. It is this second aspect I wish to discuss here.

First, it is necessary to summarize some of the accumulated results on surface critical behavior[6]. In the two-dimensional case, of course, the 'surface' is one-dimensional. For definiteness consider a semi-infinite system in the half-plane $y > 0$, close to the bulk critical temperature T_c. The system is assumed to have a single order parameter ψ, which would be the local magnetization for the Ising model. There are three generic types of boundary condition:

Free boundary condition. In this case there are no symmetry- breaking fields acting at the surface. Because surface spins have less neighbors than those in the bulk, ψ decreases from its bulk value as we approach the boundary, so that it would actually fall to zero at some distance λ outside the sample, if extrapolated. It turns out that λ remains of the same order of magnitude as the lattice spacing a, even at T_c. Thus in the continuum limit, *i.e.* on distance scales much larger than a, where conformal invariance is supposed to hold, we may assume that the boundary condition is $\psi \to 0$.

Fixed boundary condition. In this case we can imagine that the surface spins are constrained to point in the same direction. It turns out that usually this has the same effect in the continuum limit as adding a surface symmetry-breaking field. At T_c, the order parameter decays as a power y^{-x} into the bulk. Of course, this behavior must be rounded off near the surface, but the distance over which this rounding occurs is once again $O(a)$. Thus in the continuum limit, the appropriate boundary condition is $\psi \to \infty$.

Neumann boundary condition. By adjusting parameters it is possible to enforce the condition $\partial\psi/\partial n = 0$ at the boundary. It appears that this is not possible for unitary models in two dimensions, and we shall not discuss this case further here.

The important feature of all three types of boundary condition is that they contain no length scale. This is an important necessary condition for the system to be conformally invariant even in the presence of the boundary. Note that although the above three cases represent the only types of boundary condition normally considered, in situations where the bulk system is at a multicritical point, and there is more than one kind of symmetry-breaking field, other kinds of boundary condition may be contemplated. In the specific models we consider later, this feature will not arise.

To proceed further, let us limit ourselves to two dimensions, and consider the semi-infinite system to be the upper half plane $\mathrm{Im}z > 0$. It is useful to consider scaling operators as depending on complex co-ordinates $(z, \bar{z})$. In the bulk, we may always subtract any operator so that $\langle\phi(z, \bar{z})\rangle = 0$. Near the surface this is no longer true. In fact if ϕ has scaling dimension x, then $\langle\phi\rangle$ must be proportional to $(z - \bar{z})^{-x}$, since it can depend only on $\mathrm{Im}z$. In some cases the amplitude of this term may vanish for symmetry reasons, *e.g.* if ϕ is the magnetization and we are using free boundary conditions.

One important operator is the stress tensor. As in the bulk[7,8]), *local* conformal invariance implies that this has two non-zero components, which may be taken to be $T(z)$ and $\overline{T}(\bar{z})$. Since $T(z)$ does not depend on $\bar{z}$, the amplitude of its one-point function must vanish, and we conclude that $\langle T(z)\rangle = 0$, and similarly for $\langle\overline{T}\rangle$. Note that this is not in general true in other geometries. In what follows, it is important to understand the correct boundary condition on T and $\overline{T}$. The stress tensor is, as usual, defined by the response of the Hamiltonian ('action' in field theory) to an infinitesimal general co-ordinate transformation $x^\mu \to x^\mu + \alpha^\mu(x)$:

$$\delta\mathcal{A} = -\frac{1}{2\pi}\int T_{\mu\nu}\frac{\partial\alpha^\mu}{\partial x_\nu}d^2x \tag{1}$$

The semi-infinite system is invariant under conformal transformations which leave the boundary invariant. These correspond infinitesimally to analytic mappings $z \to z + \alpha(z)$ where $\overline{\alpha(z)} = \alpha(\bar{z})$. Evaluating $\delta\mathcal{A}$ in this case, we find a surface term

$$\int_{\text{boundary}} [T(z) - \overline{T}(\bar{z})]\alpha(z)dz$$

which must vanish for arbitrary $\alpha(z)$. Thus $T(z) = \overline{T}(\bar{z})$ when z is on the real axis, which means that we can think of $\overline{T}(\bar{z})$ as being the analytic continuation of $T(z)$ to the lower half plane.

The generators of the Virasoro algebra can now be defined in the usual way[2,7]:

$$\begin{aligned} L_n &= \frac{1}{2\pi i}\int_C z^{n+1}T(z)dz \\ \overline{L}_n &= -\frac{1}{2\pi i}\int_C \bar{z}^{n+1}\overline{T}(\bar{z})d\bar{z} \end{aligned} \tag{2}$$

where C is a circle surrounding the origin. The above-mentioned boundary condition then implies that $L_n = \overline{L}_n$, so that, in contrast with the usual case, we have only one Virasoro algebra. In the bulk, we know that the eigenvalues of L_0 and $\overline{L}_0$ are just the scaling dimensions $(h, \bar{h})$ of the scaling operators. It turns out that in the semi-infinite case, the eigenvalues of L_0 are the *surface* scaling dimensions $\tilde{x}$, to be defined below. In order to do this, consider the two-point function $\langle\phi(z_1,\bar{z}_1)\phi(z_2,\bar{z}_2)\rangle$. We can think of this as depending on the four complex numbers $(z_1, \ldots \bar{z}_2)$, in much the same way as does the four-point function in the bulk. Just as invariance under the $SL(2,\mathbf{C})$ subgroup fixes the form of the four-point function up to one unknown scaling function in the bulk, invariance under the group $SL(2,\mathbf{R})$ generated by L_1, L_0 and L_{-1} implies that[4]

$$\langle\phi(z_1,\bar{z}_1)\phi(z_2,\bar{z}_2)\rangle = (z_1-\bar{z}_1)^{-x}(z_2-\bar{z}_2)^{-x}\Phi\left[\frac{(z_1-z_2)(z_1-\bar{z}_2)}{(z_1-\bar{z}_1)(z_2-\bar{z}_2)}\right] \tag{3}$$

where $\Phi(\varsigma)$ is as yet undetermined. In the above, x is the scaling dimension of the operator ϕ, assumed a scalar.

As $z_1 \to z_2$, the effect of the boundary becomes unimportant, and we obtain the short-distance behavior $|z_1 - z_2|^{-2x}$. However, as $\mathrm{Re}(z_1 - z_2) \to \infty$ with $\mathrm{Im}z_1$ and $\mathrm{Im}z_2$ fixed, the two-point function is expected to fall off as $(\mathrm{Re}(z_1 - z_2))^{-2\tilde{x}}$, where, in general, $\tilde{x} > x$. This expectation comes from renormalization group calculations[6], as well as the only non-trivial exactly solved case in two dimensions, the Ising model[9]. As we shall see, the existence of $\tilde{x}$ corresponds to L_0 having a discrete spectrum. From Eq. (3) we see that $\Phi(\varsigma)$ must behave like $\varsigma^{-\tilde{x}}$ as $\varsigma \to \infty$.

The limit $\varsigma \to \infty$ corresponds to another physical region of interest. This is when $\mathrm{Im}z_1 \to 0$, *i.e.* when one of the points approaches the surface. As a result, conformal invariance completely determines the form of the correlation function between the surface and the bulk, once $\tilde{x}$ is known. This result, being based only on the 'small' conformal group, is valid also for $d > 2$.

In two dimensions, the analog between the two-point function in the semi-infinite geometry and the four-point function in the bulk goes much further. In fact in the minimal theories of Belavin, Polyakov and Zamolodchikov[7]), where the correlation functions satisfy linear partial differential equations, one may show that the n-point correlations in the surface geometry satisfy the same partial differential equations as do the $2n$-point functions in the bulk. In certain simple cases, these have been solved for the two-point functions[4]).

The physical significance of L_0 is made apparent if we make the conformal transformation $w = (l/\pi)\ln z$ to a strip of width l, with the same boundary condition. The two-point function in the strip is then

$$\langle\phi(w_1,\bar{w}_1)\phi(w_2,\bar{w}_2)\rangle_{\text{strip}} = |e^{\pi x w_1/l}||e^{\pi x w_2/l}|\langle\phi(z_1,\bar{z}_1)\phi(z_2,\bar{z}_2)\rangle_{\text{halfplane}} \quad (4)$$

where the correlation function on the right hand side is given by (3). If we now write $w = u + iv$ and consider the limit $\text{Re}(u_1 - u_2) \to \infty$, the dependence on x cancels out, and we find[10]) that the correlation function in the strip decays as $\exp(-\pi\tilde{x}(u_1 - u_2)/l)$. Now we may also think of the classical system in the strip as representing a quantum field theory defined on the interval $0 < v < l$, with u playing the role of imaginary time, in which case the exponential decay found above implies the existence of an energy gap $E_n - E_0 = \pi\tilde{x}/l$, where the E_n are the eigenvalues of the Hamiltonian $\hat{H}_F$ of the quantum field theory. This Hamiltonian is related in the usual way to the space integral of the time-time component of the stress (energy-momentum) tensor:

$$\hat{H}_F = \int_0^l T_{uu}(v)dv = \int_0^l (T(v) + \overline{T}(v))dv \quad (5)$$

On the other hand, the stress tensor in the strip is related to that in the half plane by[7])

$$T(z) = w'^2 T(w) + \frac{c}{12}\left[\frac{w'''}{w'} + \frac{3}{2}\left(\frac{w''}{w'}\right)^2\right] \quad (6)$$

From this it follows after a simple calculation that[11])

$$\hat{H}_F = \frac{\pi}{l}\left(L_0 - \frac{c}{24}\right) \quad (7)$$

which shows that the eigenvalues of L_0 are $\tilde{x}+(c/24)$. This factor of 24 is directly related to the famous 26 of string theory[12].

Now let us suppose that the model has positive Boltzmann weights. In that case, each eigenstate of L_0 which is annihilated by the L_n with $n > 0$ gives rise to a unitary highest weight representation of the Virasoro algebra. If $c < 1$, then according to Friedan, Qiu and Shenker[2] this is only possible if $c = 1-6/m(m+1)$ with $m = 2, 3, \ldots$, in which case the allowed values of the eigenvalues of L_0 are given by the Kac formula

$$\tilde{x} = \frac{[p(m+1) - qm]^2 - 1}{4m(m+1)}, \qquad (1 \leq q \leq p \leq m-1). \tag{8}$$

The important thing to note here is that the scaling dimensions x and $\tilde{x}$ of a given operator do not in general correspond to the same (p, q). Also, this result does not tell us which values of (p, q) actually appear in a given model, and to which physical operators they correspond. In the bulk, this question is answered using modular invariance of the partition function on a torus[3], and it is a generalization of that method which we shall use here[5].

Consider the theory defined on a finite cylinder, of dimensions $l \times l'$, with periodic boundary conditions in one direction, and free or fixed in the other. The partition function Z may be calculated using the Hamiltonian $\hat{H}_F$ introduced above:

$$Z = \operatorname{Tr} e^{-l'\hat{H}_F} = e^{\pi c l'/24l} \sum_{\tilde{x}} e^{-\pi \tilde{x} l'/l} \tag{9}$$

where the sum is over all surface scaling dimensions in the theory. The sum can be broken down into conformal blocks, each corresponding to a highest weight representation labelled by (p, q). The contribution from each block is given by the character formula for the representation, which may be deduced purely from the Virasoro algebra[13]. Thus, if $N_{p,q}$ is the multiplicity of the representation,

$$Z = e^{\pi c \delta/24} \sum_{p,q} N_{p,q} \chi_{p,q}(e^{-\pi\delta}) \tag{10}$$

where $\delta = l'/l$. The character formula of Ref. 13 may be written, after using the

Poisson sum formula,

$$\chi_{p,q}(z) = e^{-(\pi c/24)(\delta - 4/\delta)} \prod_{n=0}^{\infty} (1 - \tilde{z}^n)^{-1} \times \sum_{r=-\infty}^{\infty} \tilde{z}^{-(r^2-1)/m(m+1)} \sin \frac{\pi r p}{m} \sin \frac{\pi r q}{m+1} \tag{11}$$

where $\tilde{z} = e^{-\pi/\delta}$. Note that when substituted into (10) this gives a series in $e^{-1/\delta}$.

On the other hand, we can also evaluate Z by considering l, rather than l', as imaginary time, in which case the Hamiltonian $\hat{H}_P$ is that for periodic boundary conditions, whose eigenvalues are related to the bulk scaling dimensions. In this case, however, we may no longer take the trace, but instead must take an expectation value in some state $|F\rangle$ representing the boundary:

$$Z = \langle F|e^{-\hat{H}_P l}|F\rangle = e^{\pi c/6\delta} \sum_{x_n} |\langle n|F\rangle|^2 e^{-2\pi x_n/\delta} \tag{12}$$

where the sum is over the bulk scaling dimensions, presumed known.

We now compare (12) with (10), and equate coefficients of $e^{-1/\delta}$. First note the remarkable feature that the powers which occur in (12) (where x_n is given by the Kac formula) also appear in (10). Now in general we do not know much about the quantities $|\langle n|F\rangle|^2$, except that they are non-negative. For free boundary conditions, however, when $|F\rangle$ will transform according to the trivial representation of the symmetry group of the theory, this quantity must vanish whenever x_n corresponds to an operator transforming according to another representation, for example if this is the magnetization in an Ising model. It turns out that this requirement, plus the condition that the $N_{p,q}$ be non-negative integers, almost completely fixes the spectrum of $\hat{H}_F$, and hence the complete set of surface scaling dimensions.

Once the surface operator content has been found, it remains to identify the operators according to their symmetry properties. This may be done by appropriately 'twisting' the boundary conditions around the cylinder[5,14]). We now discuss the results of this somewhat laborious analysis for the critical Ising model with free boundary conditions.

First recall that the scalar operator content in the bulk case is as shown in the conformal grid below:

ϵ	$\mathbf{1}$
σ	σ
$\mathbf{1}$	ϵ

where the columns and rows are labelled by (p, q), and ϵ and σ refer to energy and magnetization operators respectively. We have extended the allowed region of (p, q) to a rectangle, so that each primary operator appears twice in the grid. Note that each primary operator transforms in a different way under the symmetries of spin flip ($\sigma \to -\sigma$, $\epsilon \to \epsilon$) and duality ($\epsilon \to -\epsilon$).

With free boundary conditions, the operator content is as follows:

σ	$\mathbf{1}$
–	–
$\mathbf{1}$	σ

Note that the magnetization operator sits where the energy operator did in the bulk. This implies that $\tilde{x}_\sigma = \frac{1}{2}$, in accord with the exact result[9]). The energy operator is no longer primary. This is because duality is no longer a symmetry. The energy operator now sits in the conformal block of $\mathbf{1}$. The most relevant operator is $L_{-2}\mathbf{1}$, with $\tilde{x} = 2$.

For the critical 3-state Potts model ($m = 5$), the bulk operator content is given by

ϵ''	ϵ'	ϵ	$\mathbf{1}$
–	–	–	–
σ'	σ	σ	σ'
–	–	–	–
$\mathbf{1}$	ϵ	ϵ'	ϵ''

With free boundary conditions, it turns out that the $N_{p,q}$ are not completely determined by the above conditions. We find two solutions, one of which has energy-like operators at (2,1) and (3,1). This solution is ruled out by the following result[15]. If ϕ denotes a primary operator whose expectation value near the boundary in the half-plane is non-zero, then its surface scaling dimension $\tilde{x}_\phi$ is an even integer.

To prove this, consider the conformal mapping[16] $w = -i(l/\pi)\arcsin(\pi z/l)$, which maps the half-plane into the semi-infinite strip $(u > 0, -l/2 < v < l/2)$. One finds that in the strip $\langle\phi\rangle$ has the form

$$\langle\phi(u,v)\rangle \propto \left[\frac{|\cosh(\pi w/l)|}{\mathrm{Re}(l/\pi)\sinh(\pi w/l)}\right]^{\tilde{x}} \tag{13}$$

Expanding this for large u/l, we get a series of integral powers of $\exp(-2\pi u/l)$. On the other hand, using the Hamiltonian $\hat{H}_F$ we find

$$\langle\phi(u,v)\rangle = \sum_n \langle F|n\rangle\langle n|\phi(v)|0\rangle e^{-\pi\tilde{x}_n u/l} \tag{14}$$

Comparing with the above, we find that $\tilde{x}$ is an even integer.

The other solution, which satisfies this constraint, is

ϵ'	–	–	**1**
–	–	–	–
σ	–	–	σ
–	–	–	–
1	–	–	ϵ'

The magnetization operators are 2-fold degenerate, as in the bulk. The low-lying eigenstates of $\hat{H}_F$ corresponding to the primary operators in the above table, plus the first few in their conformal blocks, have been found numerically by Gehlen and Rittenberg[17], confirming the above result.

So far we have discussed free boundary conditions. The fixed case is harder to treat directly since in general none of the inner products $\langle F|n\rangle$ in (12) vanish on symmetry grounds. However, for selfdual models (including the two discussed above) the surface operator content follows directly from that in the free case. This is because duality transforms a free boundary into one with an infinitely strong surface symmetry-breaking field. Explicitly, for the 3-state Potts model, if Z_F denotes the partition function for the cylinder geometry with free boundaries, and Z_{AB} denotes the same object when the opposite edges of the cylinder are fixed in the states A and B respectively (where A, B, C are the three Potts states), then

$$Z_F = Z_{AA} + Z_{AB} + Z_{AC} \tag{15}$$

where the last two are equal. It is easy to see then how the operator content of Z_F distributes itself between these terms. For Z_{AA} there remains a Z_2 symmetry between B and C. The primary operators are $\mathbf{1}$, and σ, which is odd under this symmetry. Their position in the conformal grid is as shown:

σ	–	–	$\mathbf{1}$
–	–	–	–
–	–	–	–
–	–	–	–
$\mathbf{1}$	–	–	σ

On the other hand, for the other two terms in (15) there is no symmetry left, and each has only one primary operator:

–	–	–	–
–	–	–	–
$\mathbf{1}$	–	–	$\mathbf{1}$
–	–	–	–
–	–	–	–

In this talk, I have tried to show the usefulness of the methods of conformal and modular invariance in discussing systems with boundaries. Most of the results for the surface scaling dimensions are new, although they could probably also be obtained using the Coulomb gas methods of Nienhuis[18]). We have seen that, for unitary models with $c < 1$, the surface exponents are given by the Kac formula. However, there are generally less primary operators than in the bulk. This number decreases as we enforce boundary conditions which break more and more of the symmetry.

This work was supported by NSF Grant No. PHY83-13324.

REFERENCES

1) See, for example, Kleban P., Akinci G., Hentschke R. and Brownstein K.R., *J. Phys.* **A19**, 437, (1986); Bartelt N.C. and Einstein T.L., *J.Phys.***A**, to appear.

2) Freidan D., Qiu Z. and Shenker S., *Phys. Rev. Lett.* **52**, 1575, (1984).

3) Cardy J.L., *Nucl. Phys.* **B270**, 186, (1986).

4) Cardy J.L., *Nucl. Phys.* **B240**, 514, (1984).

5) Cardy J.L., *Nucl. Phys.* **B275**, 200, (1986).

6) Dietrich S. in *Phase Transitions and Critical Phenomena*, v.10, C.Domb & J.Lebowitz eds., (Academic,London,1986).

7) Belavin A.A., Polyakov A.M. and Zamolodchikov A.B., *Nucl. Phys.* **B241**, 333, (1984).

8) Cardy J.L. in *Phase Transitions and Critical Phenomena*, v.11, C.Domb & J.Lebowitz eds., (Academic,London,1986).

9) McCoy B. and Wu T.T., *Phys. Rev.* **162**, 436, (1967).

10) Cardy J.L., *J. Phys.* **A17**, L385, (1984).

11) Blöte H.W.J., Cardy J.L. and Nightingale M.P., *Phys. Rev. Lett.* **56**, 742, (1986).

12) Brink L. and Nielsen H.B., *Phys. Lett.* **43B**, 319, (1973).

13) Rocha-Caridi A. in *Vertex Operators in Mathematics and Physics*, ed. J.Lepowsky (Springer, New York, 1984), p.451.

14) Zuber J.-B., *Phys. Lett.* **176B**, 127, (1986).

15) Burkhardt T.W. and Cardy J.L., in preparation.

16) Burkhardt T.W. and Eisenriegler E., *J. Phys.* **A18**, L25, (1985).

17) Gehlen G.v. and Rittenberg V., *J.Phys* **A**, to appear.

18) Nienhuis B. in *Phase Transitions and Critical Phenomena*, v.11, C.Domb & J.Lebowitz eds., (Academic,London,1986)., and in these proceedings.

INTEGRABLE THEORIES AND CONFORMAL INVARIANCE

H.J. de Vega*)
Theoretical Physics Division, CERN
1211 Geneva 23, Switzerland

ABSTRACT

Gapless integrable theories in two dimensions exhibit conformal invariance for long distances. After a brief summary of the Bethe ansatz method, a general procedure to compute the conformal characteristics (central charge and scaling dimensions) is exposed. The six-vertex model is treated in detail.

1. INTRODUCTION

Theories possessing as many commuting and conserved physical magnitudes as degrees of freedom are called integrable. So a field theory or a statistical model must have an infinite number of conserved quantities in order to be integrable.

Exact eigenvalues and eigenstates of the infinite set of commuting operators that includes the Hamiltonian and the momentum are often explicitly calculable. In this way the exact mass spectrum and S-matrix in field theory and the free energy in statistical mechanics follow. This is achieved using the Bethe ansatz and its generalizations mainly in the modern language of the quantum inverse method[1),2)].

---

*) Permanent address: LPTHE, Université Paris VI, Tour 16, 1er étage, 4, Place Jussieu, 75230 Paris, Cedex 05, France

Order parameters[2], form factors and correlation functions are also calculable to some extent[3].

Integrable statistical models are often gapless or possess regimes in the parameter space where the mass gap vanishes. For example, the six-vertex model is gapless for $|\Delta| < 1$. It corresponds to the critical regime of the more general eight-vertex model. Multistate vertex models also possess zero gap and non-zero gap regimes usually separated by a point of higher symmetry. These scale-invariant integrable models are then good candidates to exhibit two-dimensional conformal invariance for distances much larger than the lattice spacing. In fact, they possess a much richer structure than that associated with the conformal invariance since they are exactly solvable.

A simple way to obtain the conformal properties of a model is to look at its finite size properties for large size[4]. Let E_n be the eigenvalues of the Hamiltonian on an interval on length L and periodic boundary conditions. Then

$$E_0 = fL - \frac{\pi c}{6L} \tag{1.1}$$

$$E_n - E_0 = \frac{2\pi x_n}{L} \tag{1.2}$$

Here E_0 stands for the ground state energy, c the value of the central charge and x_n the scaling dimension of the operators associated to the excited state E_n.

In order to obtain c and the x_n for integrable theories, it is then enough to know the eigenvalues E_n for large but finite size. However, one finds in the literature[1),2)] explicit results only at $L = \infty$. That is f in Eq. (1.1). In Ref. 5) a method to compute finite size effects in integrable models was proposed and applied to some models. Using this method, one can compute finite size corrections both for non-zero gap and gapless regimes. In the first case, one finds exponentially small corrections in L[5),6)]. We shall discuss in this lecture the computation of finite size correction for gapless models following a recent work in collaboration with M. Karowski[7].

In fact, there are two ways of computing finite size corrections in a two-dimensional system. The partition function reads at temperature T

$$Z = \mathrm{Tr}\left(e^{-\beta H}\right) \tag{1.3}$$

where $\beta = 1/T$ and H is considered in an interval of length L with periodic boundary conditions. Since all physical magnitudes are periodic in the imaginary time with period $2\pi\beta$, one can get information about finite (but large) size either from

A) $L = \infty$, β large but finite

or

B) $T = 0$ and L large and finite.

I shall restrict myself here to method B). Method A) can be found in Refs. 7) and 8).

I expose in this lecture the results for the six-vertex model and models related to it, like the critical Potts model. In the same way it is possible to derive the conformal properties of multistate integrable theories[9].

For the six-vertex model, it is shown here that the central charge equals one. For the critical q-state Potts model, we find[7]

$$c = 1 - \frac{6}{\nu(\nu-1)}$$

where $\sqrt{q} = 2\cos\pi/\nu$, $\nu = 3,4,\ldots$. This exact result is in agreement with previous predictions[4),10)].

In Section 2, I summarize the exact solution of the six-vertex model by Bethe ansatz. In Section 3, the method to calculate finite size corrections in integrable theories[5] is exposed as well as its applications[7].

2. THE SIX-VERTEX MODEL

I recall in this section the definition and the construction of eigenvectors and eigenvalues of the six-vertex model.

Let us consider a two-dimensional lattice of size N×M where each bond can be in two different states (1 and 2). The statistical weight of a vertex configuration depends on the state of the four bonds joining at the vertex. We write this weight as usual

$$[t_{ab}(\theta)]_{ij}$$

Here a, b stand for the horizontal links and i,j for the vertical ones. Then, one associates a transfer operator or monodromy operator to a horizontal line in the lattice as follows:

$$T_{ab}(\theta) = \sum_{a_1 \cdots a_{N-1}} t_{aa_1}(\theta) \otimes t_{a_1 a_2}(\theta) \otimes \cdots \otimes t_{a_{N-1} b}(\theta) \tag{2.1}$$

Here the tensor product concerns the vertical two-dimensional spaces. The six-vertex model is defined by

$$t_{11}(\theta) = \begin{pmatrix} \sin(\theta+\gamma) & 0 \\ 0 & \sin\theta \end{pmatrix}, \quad t_{22}(\theta) = \begin{pmatrix} \sin\theta & 0 \\ 0 & \sin(\theta+\gamma) \end{pmatrix}$$

$$t_{12} = \sigma_- \sin\gamma \quad , \quad t_{21} = \sigma_+ \sin\gamma \tag{2.2}$$

in the $|\Delta| < 1$ regime. For $\Delta < -1$ one has analogous expressions with hyperbolic functions instead of trigonometric ones. It follows that $T_{ab}(\theta)$ obeys a Yang-Baxter algebra:

$$R(\theta-\theta')[T(\theta) \otimes T(\theta')] = [T(\theta') \otimes T(\theta)]R(\theta-\theta') \tag{2.3}$$

with R-matrix

$$R(\theta) = \begin{pmatrix} \sin(\theta+\gamma) & 0 & 0 & 0 \\ 0 & \sin\gamma & \sin\theta & 0 \\ 0 & \sin\theta & \sin\gamma & 0 \\ 0 & 0 & 0 & \sin(\theta+\gamma) \end{pmatrix} \tag{2.4}$$

As usual we set

$$T(\theta) = \begin{pmatrix} A(\theta) & B(\theta) \\ C(\theta) & D(\theta) \end{pmatrix} \tag{2.5}$$

One gets a family of commuting transfer matrices

$$\tau(\theta) = A(\theta) + D(\theta) \tag{2.6}$$

from the Yang-Baxter algebra (2.3). However, other families of commuting transfer matrices exist besides (2.6)[11].

The eigenvectors of $\tau(\theta)$ are easily constructed by the algebraic Bethe ansatz[12]. One chooses a reference state to start

$$|\Omega\rangle = \begin{pmatrix} 1 \\ 0 \end{pmatrix}_{(1)} \otimes \cdots \otimes \begin{pmatrix} 1 \\ 0 \end{pmatrix}_{(N)} \tag{2.7}$$

This is an eigenstate of $A(\theta)$ and $D(\theta)$

$$\begin{aligned} A(\theta)\,|\Omega\rangle &= \sin^N(\theta+\gamma)\,|\Omega\rangle \\ D(u)\,|\Omega\rangle &= \sin^N\theta\,|\Omega\rangle \end{aligned} \tag{2.8}$$

The one looks for eigenstates of $\tau(\theta)$ with the form

$$\Psi(\theta_1,\ldots,\theta_r) = \prod_{j=1}^{r} B(\theta_j)\,|\Omega\rangle \tag{2.9}$$

The number $\theta_1,\ldots,\theta_r$ will be determined below by the requirement that ψ is an eigenstate of $\tau(\theta)$. Using the Yang-Baxter algebra as defined by Eqs. (2.3) and (2.4), one can compute the action of $A(\theta)$ and $D(\theta)$ on the states (2.9). One finds

$$A(\theta)\,\Psi(\theta_j) = \Lambda_+(\theta,\theta_j)\,\Psi(\theta_j) + \sum_{s=1}^{r} \Lambda_s(\theta,\theta_j)\,\Psi_s(\theta_j)$$

$$D(\theta)\,\Psi(\theta_j) = \Lambda_-(\theta,\theta_j)\,\Psi(\theta_j) + \sum_{s=1}^{r} \Lambda'_s(\theta,\theta_j)\,\Psi_s(\theta_j) \tag{2.10}$$

where we used also Eq. (2.8) and we introduce the notation

$$\Lambda_+(\theta,\theta_j) = \sin^N(\theta+\gamma) \prod_{j=1}^{r} \frac{\sin(\theta+i\lambda_j-\gamma/2)}{\sin(\theta+i\lambda_j+\gamma/2)} \quad (-1)^r \tag{2.11}$$

$$\Lambda_-(\theta,\theta_j) = \sin^N\theta \prod_{j=1}^{r} \frac{\sin(\theta+i\lambda_j+\frac{3}{2}\gamma)}{\sin(\theta+i\lambda_j+\frac{\gamma}{2})} \quad (-1)^r$$

$$\Lambda_s(\theta,\theta_j) = \frac{\sin\gamma\,\sin^N(\gamma+i\lambda_s)}{\sin(\theta+i\lambda_s)} \prod_{j\neq s} \frac{\sinh(\lambda_j-\lambda_s+i\gamma)}{\sinh(\lambda_j-\lambda_s)}$$

$$\Lambda'_s(\theta,\theta_j) = -\frac{\sin\gamma\,\sin^N(i\lambda_s)}{\sin(\theta+i\lambda_s)} \prod_{j\neq s} \frac{\sinh(\lambda_s-\lambda_j+i\gamma)}{\sinh(\lambda_s-\lambda_j)} \tag{2.12}$$

Here $\lambda_j = \theta_j + i\gamma/2$. These formulae are written for $|\Delta| < 1$. Analogous expressions hold for $|\Delta| > 1$.

As it is clear from Eq. (2.10), ψ will be an eigenstate of $A(\theta) + D(\theta) = \tau(\theta)$ when

$$\Lambda_s + \Lambda'_s = 0 \quad , \quad 1 \leq s \leq r$$

More explicitly, the Bethe ansatz equations (BAE) read

$$\left[\frac{\sinh(\lambda_s + i\gamma/2)}{\sinh(\lambda_s - i\gamma/2)}\right]^N = -\prod_{j=1}^{r} \frac{\sinh(\lambda_s - \lambda_j + i\gamma)}{\sinh(\lambda_s - \lambda_j - i\gamma)} \tag{2.13}$$

Here the θ-dependence has cancelled as it must. (Since $[\tau(\theta),\tau(\theta')] = 0$, the eigenvectors of $\tau(\theta)$ can be chosen θ-independent.) Now after solving Eq. (2.13) for $(\lambda_1,\ldots,\lambda_r)$, one obtains the corresponding eigenvalue of $\tau(\theta)$

$$\Lambda(\theta,\lambda_j) = \Lambda_+(\theta,\lambda_j) + \Lambda_-(\theta,\lambda_j) \tag{2.14}$$

by inserting the roots $(\lambda_1,\ldots,\lambda_r)$ in Eqs. (2.11).

As it is well known, the XXZ Hamiltonian is generated by the six-vertex transfer matrix as

$$H_{XXZ} = -\sqrt{|\Delta^2-1|}\,\frac{\partial}{\partial\theta}\log\tau(\theta)\Big|_{\theta=0} \tag{2.15}$$

So, the eigenvectors (2.9) are also eigenstates of H_{XXZ} and its eigenvalues follow from Eqs. (2.11)-(2.15). The Bethe ansatz construction (2.9)-(2.13) admits various generalizations for multistate vertex models[11),13)] (q states per link instead of two) and for higher spin generalization of the XXZ Heisenberg Hamiltonian[14)].

Another very interesting generalization consists in introducing a local shift $\theta \to \theta + \mu_a$ $(1 \leq a \leq N)$ in the local weights[15),11)] [see, e.g., (2.1)]. These shifts respect the YB algebra and hence the integrability. It has been recently shown that all these integrable lattice models possess a hidden local gauge symmetry[16)]. Physical observables turn to be gauge invariant and the gauge group may be Abelian or non-Abelian. It is a local U(1) symmetry for the six-vertex model and the XXZ Hamiltonian.

3. FINITE SIZE CALCULATIONS AND CONFORMAL INVARIANCE

We analyze in this section the BAE for the six-vertex model derived in Section 2. Taking the logarithm Eq. (2.13) yields for real λ_i:

$$N\phi(\lambda_i, \frac{\gamma}{2}) = \sum_{j=1}^{r} \phi(\lambda_i - \lambda_j, \gamma) + 2\pi I_i \qquad 1 \le i \le r \tag{3.1}$$

Here for $\Delta < -1$, $\Delta = -\text{ch}\gamma$ and

$$\phi(z,\alpha) = i \,\text{Log} \frac{\sin(z+i\alpha)}{\sin(z-i\alpha)} \tag{3.2}$$

At $\Delta = -1$

$$\phi(z,\alpha) = i \,\text{Log} \frac{z+i\alpha}{z-i\alpha}$$

and for $|\Delta| < 1$, we set $\Delta = \cos\gamma$ and

$$\phi(z,\alpha) = i \,\text{Log} \frac{\sinh(z+i\alpha)}{\sinh(z-i\alpha)} \tag{3.3}$$

In both cases the cut of the logarithm is taken such $\phi(x,\alpha)$ is a continuous function for real $x \in (-A,A)$. We have $A = \pi/2$ when $\Delta < -1$ and $A = \infty$ when $-1 \leq \Delta < 1$. In Eq. (3.1) the I_i are half-odd integers and we take N to be even.

The eigenvalues of the transfer matrix read from Eq. (2.12)

$$\Lambda(\theta) = \Lambda_+(\theta) + \Lambda_-(\theta)$$

where

$$\text{Log}\,\Lambda_+(\theta, \lambda_j) = -i \sum_{j=1}^{r} \phi(\lambda_j - i\theta, \gamma/2)$$

$$\text{Log}\,\Lambda_-(\theta, \lambda_j) = i\phi(i\theta + i\frac{\gamma}{2}, \frac{\gamma}{2}) + i\sum_{j=1}^{r} \phi(\lambda_j - i\theta - i\gamma, \frac{\gamma}{2}) \tag{3.4}$$

Here we have rescaled all statistical weights by a factor $\sin(\theta+\gamma)$

$[\mathrm{sh}(\theta+\gamma)]$ for $|\Delta| < 1$ $[(\Delta < -1)]$. After Eqs. (2.15) and (3.4), the energy eigenvalues per site of the XXZ Hamiltonian write

$$E_N = -\frac{\sqrt{|\Delta^2-1|}}{N} \sum_{j=1}^{r} \phi'(\lambda_j, \tfrac{\gamma}{2}) \tag{3.5}$$

In order to analyze the BAE for finite N, it is convenient to use the function[5)]

$$z_N(\lambda) = \frac{1}{2\pi}\left[\phi(\lambda, \tfrac{\gamma}{2}) - \frac{1}{N}\sum_{i=1}^{r}\phi(\lambda-\lambda_i, \gamma)\right] \tag{3.6}$$

This function is continuous and monotonically increasing for real λ. At the real roots of the BAE

$$z_N(\lambda_i) = \frac{I_i}{N} \tag{3.7}$$

It is also useful to consider its derivative

$$\sigma_N(\lambda) = \frac{dz_N}{d\lambda} \tag{3.8}$$

The half-odd integers I_i form an equally spaced monotonic sequence for the ground state

$$I_{j+1} - I_j = 1 \tag{3.9}$$

For excited states the sequence exhibits jumps for some values of i

$$I_{i+1} - I_i = 1 + \sum_{h=1}^{N_h} \delta_{i i_h} \tag{3.10}$$

The values of λ associated with these missing half-integers are called holes and denoted by θ_h

$$z_N(\theta_h) = \frac{I_{i_h}+1}{N} \tag{3.11}$$

Each jump in the sequence corresponds to removing a root from the ground state at the point θ_h.

When N goes to infinity the λ_j tend to have a continuum distribution with density

$$\rho_\infty(\lambda_j) = \lim_{N\to\infty} \frac{1}{N(\lambda_{j+1}-\lambda_j)} \tag{3.12}$$

One finds from Eqs. (3.7) and (3.8) for large N

$$\sigma_\infty(\lambda_j) = \lim_{N\to\infty} \frac{I_{j+1}-I_j}{N(\lambda_{j+1}-\lambda_j)} \tag{3.13}$$

Then Eqs. (3.11)-(3.13) yield for large N

$$\sigma_\infty(\lambda) = \rho_\infty(\lambda) + \frac{1}{N}\sum_{h=1}^{N_h} \delta(\lambda - \theta_h) \tag{3.14}$$

A linear integral equation for $\sigma_\infty(\lambda)$ follows by taking the difference between Eqs. (3.1) for two adjacents real roots λ_{k+1} and λ_k:

$$\sigma_\infty(\lambda) = \frac{1}{2\pi}\phi'(\lambda,\frac{\gamma}{2}) - \int_{-A}^{A}\frac{d\mu}{2\pi}\sigma_\infty(\mu)\,\phi'(\lambda-\mu,\gamma) + \\ + \frac{1}{2\pi N}\sum_h \phi'(\lambda-\theta_h,\gamma) - \frac{1}{2\pi N}\sum_\ell\left[\phi'(\lambda-\xi_\ell,\gamma)+\phi'(\lambda-\bar{\xi}_\ell,\gamma)\right] \tag{3.15}$$

The following relation was used to derived Eq. (3.15):

$$\lim_{N\to\infty}\frac{1}{N}\sum_j f(\lambda_j) = \int_{-A}^{A} f(\lambda)\,\rho_\infty(\lambda)\,d\lambda \tag{3.16}$$

We denoted in (3.15) by ξ_ℓ, $\bar{\xi}_\ell$ the complex roots ($\text{Im}\,\xi_\ell > 0$). They always appear in conjugate pairs. Fourier expansions solve this equation. The solution reads

$$\sigma_\infty(\lambda) = \sigma_\infty^v(\lambda) + \frac{1}{N}\left[\sigma_h(\lambda) + \sigma_{\mathbb{C}}(\lambda)\right]$$

$\sigma_\infty^v(\lambda)$ corresponds to the ground state. One finds

$$\sigma_\infty^v(\lambda) = \frac{1}{2\pi}\sum_{m\in\mathbb{Z}}\frac{e^{2mi\lambda}}{\cosh m\lambda} = \frac{K(k)}{\pi^2}\,dn\left(\frac{2K\lambda}{\pi},k\right) \quad , \Delta<-1 \tag{3.17}$$

(Here $K(k)/K'(k) = \gamma/\pi$) ;

$$\sigma_\infty^V(\lambda) = \int_{-\infty}^{+\infty} \frac{dk}{4\pi} \frac{e^{ik\lambda}}{\cosh \frac{k\gamma}{2}} = \frac{1}{2\gamma \cosh \frac{\pi\lambda}{\gamma}} \,, \quad -1<\Delta<1 ;$$

$$\sigma_\infty^V(\lambda) = \frac{1}{2\cosh \pi\lambda} \,, \quad \Delta = -1 .$$

$\sigma_h(\lambda)$ stands for the hole contribution to the density of real roots

$$\sigma_h(\lambda) = \frac{1}{\pi} \sum_h p(\lambda - \theta_h) \tag{3.18}$$

where

$$p(\lambda) = \frac{1}{2} + 2 \sum_{m=1}^{\infty} \frac{\cos 2m\lambda}{e^{2m\gamma}+1} \,, \quad \Delta < -1 ;$$

$$p(\lambda) = \frac{1}{2} \int_0^\infty \frac{\cos k\lambda \ \mathrm{sh}\,(\frac{\pi}{2}-\gamma)k}{\mathrm{sh}\left[\frac{\pi-\gamma}{2}k\right] \cosh\left(\frac{k\gamma}{2}\right)} dk \quad -1<\Delta \tag{3.19}$$

$$p(\lambda) = \frac{1}{2} \int_0^\infty \frac{\cos k\lambda}{\cosh \frac{k}{2}} e^{-k/2} dk \,, \quad \Delta = -1 .$$

$\sigma_c(\lambda)$ stands for the complex roots contribution[5),16)].

The densities (3.16)-(3.19) allow an easy computation of the eigenvalues of $\tau(\theta)$ and H_{XXZ} in the infinite N limit using Eq. (3.16). In this lecture I will concentrate on finite N corrections to these eigenvalues[5)] and their connections to the central charge of the Virasoro algebra[7)]. Let us consider $\Lambda_+(\theta)$ since it dominates the free energy for large N. Define

$$L_N(\theta) = \frac{1}{N} \log \Lambda_+(\theta) - \lim_{N\to\infty} \frac{1}{N} \log \Lambda_+(\theta) \tag{3.20}$$

Obviously, $L_\infty(\theta) = 0$. We want to derive the dominant behaviour of $L_N(\theta)$ for large N. Equations (3.4), (3.14) and (3.16) yield

$$L_N(\theta) = -i\int_{-A}^{A} d\lambda\, \phi\left(\lambda - i\theta, \frac{\gamma}{2}\right)\left\{\frac{1}{N}\sum_{k=1}\delta(\lambda-\lambda_k) + \frac{1}{N}\sum_{h}\delta(\lambda-\Theta_h) - \sigma_N(\lambda)\right\} + i\int d\lambda\,[\sigma_N(\lambda)-\sigma_\infty(\lambda)]\,\phi\left(\lambda+i\theta, \frac{\gamma}{2}\right) \quad (3.21)$$

Here, λ_i ($1 \leq i \leq M$) are the real roots of Eq. (3.1). We now study the difference $\sigma_N(\lambda) - \sigma_\infty(\lambda)$. One finds from Eqs. (3.6), (3.8) and (3.15)

$$\sigma_N(\lambda) - \sigma_\infty(\lambda) + \int \frac{d\mu}{2\pi}\phi'(\lambda-\mu,\gamma)\,[\sigma_N(\mu)-\sigma_\infty(\mu)] =$$

$$= -\int \frac{d\mu}{2\pi}\phi'(\lambda-\mu,\gamma)\left\{\frac{1}{N}\sum_{i=1}^{M}\delta(\mu-\lambda_i) + \frac{1}{N}\sum_{h=1}^{N_h}\delta(\mu-\Theta_h) - \sigma_N(\mu)\right\} \quad (3.22)$$

This can be considered as a linear integral equation for $\sigma_N(\lambda) - \sigma_\infty(\lambda)$ with the right-hand side as an inhomogeneous term. Solving it yields

$$\sigma_N(\lambda) - \sigma_\infty(\lambda) = -\int_{-A}^{A}\frac{d\mu}{\pi}P(\lambda-\mu)\left\{\frac{1}{N}\sum_{i=1}^{M}\delta(\mu-\lambda_i) + \frac{1}{N}\sum_{h=1}^{N_h}\delta(\mu-\Theta_h) - \sigma_N(\mu)\right\} \quad (3.23)$$

An analogous expression for $L_N(\theta)$ follows by inserting Eq. (3.23) in Eq. (3.21)

$$L_N(\theta) = i\int_{-A}^{A} d\lambda\, F_\gamma(\lambda)\left\{\frac{1}{N}\sum_{j=1}^{M}\delta(\lambda-\lambda_j) + \frac{1}{N}\sum_{h=1}^{N_h}\delta(\lambda-\Theta_h) - \sigma_N(\lambda)\right\} \quad (3.24)$$

Here
$$F_\gamma(\lambda) = \frac{\pi\gamma}{2(\pi-\gamma)} + 2\,\mathrm{arctg}\, e^{\frac{\pi}{\gamma}(\lambda+i\theta)}, \quad |\Delta| < 1 .$$

and

$$F_\gamma(\lambda) = am(\lambda + i\theta, k) - \frac{\pi}{2} + 2\int_0^{\lambda+\pi/2} \rho(\mu)\,d\mu\,, \quad \Delta < -1\,. \tag{3.25}$$

We arrive now to the crucial step in the calculation of large N effects. We need to evaluate for large N expressions like

$$I_N = \int_{-A}^{A} d\lambda\, f(\lambda) \left\{ \frac{1}{N}\sum_{i=1}^{M} \delta(\lambda-\lambda_i) + \frac{1}{N}\sum_{h=1}^{N_h} \delta(\lambda-\theta_h) - \sigma_N(\lambda) \right\} \tag{3.26}$$

where $f(\lambda)$ is explicitly known.

It is convenient to change in (3.26) to the intergration variable $z_N(\lambda)$ as defined by (3.6). Using Eq. (3.7) and (3.11) yields

$$I_N = \int_0^P f(\lambda_N(z))\,dz \left\{ \frac{1}{N}\sum_{k=1}^{M+N_h} \delta(z-z_k) - 1 \right\} \tag{3.27}$$

where

$$P = \int_{-A}^{A} \sigma_N(\lambda)\,d\lambda = \frac{M+N_h}{N}$$

and we choose

$$z_k = \frac{k+1/2}{N}\,, \quad 1 \le k \le M+N_h \tag{3.28}$$

By Fourier expanding the periodic $\delta(z)$ with period P, one gets

$$\frac{1}{N}\sum_{k=1}^{M+N_h} \delta(z-z_k) = \sum_{\alpha=-\infty}^{+\infty} (-1)^\alpha e^{2\pi i N\alpha z} \tag{3.29}$$

Inserting Eq. (3.29) in Eq. (3.28) gives[5]

$$I_N = \sum_{\substack{\alpha=-\infty \\ \alpha\neq 0}}^{+\infty} (-1)^\alpha\, T_{N\alpha} \tag{3.30}$$

where

$$T_n = \int_{-A}^{A} f(\lambda)\, \sigma_N(\lambda)\, e^{2\pi i n z_N(\lambda)}\, d\lambda \tag{3.31}$$

Expressions (3.30) and (3.31) are exact for all values of N. Now one can proceed to obtain their asymptotic behaviour for large N. The dominant large N behaviour follows by replacing

$$T_n \simeq T_n^{as} \equiv \int_{-A}^{A} f(\lambda)\, \sigma_\infty(\lambda)\, e^{2\pi i n z_\infty(\lambda)}\, d\lambda \tag{3.32}$$

The right-hand side of T_n^{as} is known exactly from Eqs. (3.17) and (3.8)

$$z_\infty(\lambda) = \frac{1}{2\pi i} \log\left[\mathrm{sn}\left(\frac{2K\lambda}{\pi}\right) - i\,\mathrm{cn}\left(\frac{2K\lambda}{\pi}\right)\right] \tag{3.33}$$

$$z_\infty(\lambda) = \frac{1}{\pi} \operatorname{arctg} e^{\frac{\pi\lambda}{\gamma}} \tag{3.34}$$

The regimes $\Delta < -1$ and $-1 < \Delta < 1$ must be treated separately. In the first case (non-zero gap) T_n^{as} (and T_n) are dominated for large n by a complex saddle point at

$$\lambda = \lambda_0 = \frac{\pi}{2} + \frac{i\gamma}{2} \bmod (i\pi, \gamma) \tag{3.35}$$

with $\quad 2\pi i\, z_\infty(\lambda_0) = \ln \sqrt{k_1}$

Here $k_1 = [(1-k')/k]^2 < 1$. Then one finds exponentially small contributions to T_n^{as} of order $k_1^{|n|/2}$. Detailed results can be found in Ref. 5). In conclusion, the terms with $\alpha = \pm 1$ dominate the sum (3.30) and

$$I_N = O(k_1^{N/2}) \quad \text{for } N \gg 1 . \tag{3.36}$$

In the second case (gapless, $|\Delta| \leqslant 1$), the saddle point is at $|\lambda| = +\infty$

and so T_n is dominated for large n by the endpoints of integration. It is convenient to rewrite T_n^{as} as

$$T_n^{as} = \int_0^{\pi} \frac{dt}{2\pi}\, e^{int}\, f(\lambda(t)) \tag{3.37}$$

where

$$\lambda(t) = \frac{\gamma}{\pi} \ln\left[tg \frac{t}{2} \right]$$

as follows from Eq. (3.34). A typical behaviour of $f(\lambda)$ for large λ is

$$f(\lambda) \underset{\lambda \to \pm\infty}{=} f_{\pm} + g_{\pm}\, e^{-\frac{\pi}{\gamma}|\lambda|} + \text{smaller terms} \tag{3.38}$$

[see, for example Eq. (3.25)]. This gives

$$T_n^{as} \underset{n\to\infty}{=} \frac{f_+ - f_-}{2\pi n i} - \frac{g_+ + g_-}{\pi (2n)^2} + O\left(\frac{1}{n^3}\right) \tag{3.39}$$

where n is assumed even. Inserting Eq. (3.39) in (3.30) and summing over α yields

$$I_N \underset{N\gg 1}{=} -\frac{\pi}{24N^2}(g_+ + g_-) + O\left(\frac{1}{N^4}\right) \tag{3.40}$$

Let us now apply this result to the free energy correction $L_N(\theta)$ [Eq. (3.24)]. One finds comparing Eq. (3.25) and (3.38)

$$g_{\pm} = \mp 2 e^{\mp \frac{i\pi\theta}{\gamma}}$$

Then,

$$L_N(\theta) \underset{N\gg 1}{=} -\frac{\pi}{6N^2} \sin\frac{\pi\theta}{\gamma} + O\left(\frac{1}{N^4}\right) \tag{3.41}$$

So, this is the leading finite size correction to the six-vertex free energy for periodic boundary conditions in the gapless regime ($-1 < \Delta < 1$). Antiperiodic boundary conditions can be also considered.

As it was stated in Section 2, a conformally invariant field theory has a leading finite size connection equal to

$$-\frac{\pi c}{6N^2} \tag{3.42}$$

However, one cannot blindy identify Eqs. (3.41) and (3.42). Unless $|\theta| = +\gamma/2$ the model is not invariant under a $\pi/2$ rotation as one sees from the vertex weights (2.2). [One must also use the fact that change of sign of the weights leave the free energy invariant[2].] In that case $\sin\pi\theta/\gamma = 1$ and we read $c = 1$ from Eq. (3.41).

For large distances one expects that rotational invariance will be restored for all values of θ since the model is critical provided $|\Delta| < 1$. Let us analyze the elementary excitations: they are the holes and the strings. The eigenvalue of $\tau(\theta)$ reads for a hole at $\theta_{hole} = \phi$

$$\lambda(\theta,\varphi) = 2i\,\mathrm{arctg}\, e^{\frac{\pi}{\gamma}(\varphi+i\theta)} \tag{3.43}$$

Since the momentum is given here by

$$p = -i \log \tau(0)$$

one finds for this hole

$$p = 2\,\mathrm{arctg}\, e^{\pi\varphi/\gamma} \tag{3.44}$$

In this context the Hamiltonian can be identified with

$$H = -\mathrm{Re}\,\mathrm{Log}\,\tau(\theta)$$

So, we find for low energy and momentum

$$\begin{aligned} p \underset{\varphi\to-\infty}{=} & \; 2e^{\pi\varphi/\gamma} + o(e^{2\pi\varphi/\gamma}) \\ E = & \; p \sin\frac{\pi\theta}{\gamma} \end{aligned} \tag{3.45}$$

This indicates that we must renormalize the energy by a factor

$\mathrm{cosec}(\pi\theta/\gamma)$ in order to have an ultra-relativistic dispersion law and hence rotational invariance for large distances. After this renormalization

$$L_N \quad \tilde{L}_N(\theta) = \frac{1}{\sin\frac{\pi\theta}{\gamma}} L_N(\theta) = \frac{\pi}{6N^2} + o\left(\frac{1}{N^4}\right) \tag{3.46}$$

And one sees that $c = 1$ for all values of θ and $|\Delta| < 1$.

We compute here the dominant finite size contribution. Higher order terms can also be computed[5),18)]. It is also possible to compute scaling dimensions from the Bethe ansatz equations combining the methods exposed in the present section and Ref. 19)[20)]. In addition, generalization to multistate vertex models solvable by nested Bethe ansätze are possible[9)].

4. EXCITED STATES AND SCALING DIMENSIONS

Let us now consider excited states in the six-vertex model. Finite size corrections to their energies give the scaling dimensions of the operator associated to them through Eq. (1.2)

Let us consider a state with $r = N/2 - 1$. For large N, one can build a state like that assuming that the roots of the BAE are real and restricted to a finite but large interval $(-b,+b)$. In this way one can take $N = \infty$ and finds for the density of roots[19)]

$$\sigma(\lambda,y) = \frac{1}{2\pi}\phi'\left(\lambda,\frac{\gamma}{2}\right) + \int_{-b}^{b}\frac{d\mu}{2\pi}\,\sigma(\mu,y)\,\phi'(\lambda-\mu,\gamma) \tag{4.1}$$

where y is the magnetization

$$y = 1 - \frac{2r}{N} \qquad \text{and} \qquad b = b(y)$$

For $r = N/2 - 1$ one finds $y = 2/N \ll 1$ so Eq. (4.1) can be solved approximately using Wiener-Hopf techniques. Define

$$L_y(\theta) = -\lim_{N\to\infty} \frac{1}{N}\left[\log \Lambda_+(\theta,y) - \log \Lambda_+(\theta,0)\right] \tag{4.2}$$

Subtracting Eq. (3.15) from Eq. (4.1), one derives that

$$\sigma(\lambda,y) - \sigma(\lambda,0) = \int_{b(y)<|\mu|<\infty} \frac{d\mu}{\pi}\, \rho(\lambda-\mu)\, \sigma(\mu,y) \tag{4.3}$$

where $\sigma(\lambda,0) \equiv \sigma_\infty(\lambda)$. Moreover, it can be shown from Eqs. (4.2) and (4.3) that

$$L_y(\theta) = \frac{i\gamma y}{2} + 2\pi i \int_{|\lambda|>b(y)} d\lambda\, \sigma(\lambda,y)\, z_\infty(\lambda - i\theta) \tag{4.4}$$

where Eq. (3.19) was used and $z_\infty(\lambda)$ is given by Eq. (3.34). Using the perturbative solution of Eq. (4.4) given in Ref. 19), one finds

$$L_y(\theta) = \frac{y^2}{4}(\pi-\gamma)\sin\frac{\pi\theta}{\gamma}\left[1 + O(y^2) + O\left(y^{\frac{4\gamma}{\pi-\gamma}}\right)\right] \tag{4.5}$$

In the calculation we drop a term $-iy\pi/2$ since this represents a contribution to the partition function equal to one. Now, setting $y = 2/N$ and renormalizing $L_y(\theta)$ according to Eq. (3.46) yields

$$\tilde{L}_y(\theta) = \frac{\pi-\gamma}{N^2} \tag{4.6}$$

According to conformal field theory arguments [Eq. (1.2)], this corresponds to an operator of conformal dimension

$$x = \frac{1}{2}\left(1 - \frac{\gamma}{\pi}\right) \tag{4.7}$$

It can be identified to a transverse field $\Sigma_n \sigma^x_n$. This corresponds in the eight-vertex model to an "electric field"[20].

REFERENCES

1) See for reviews:
a) Faddeev, L.D., Soviet Sci. Review C1, 107 (1980) and Les Houches Lectures, North Holland (1980);
b) Kulish, P.P. and Sklyanin, E.K., in Tvärminne Lectures, Springer Lectures in Physics, Vol. 151 (1981);
c) Thacker, H.B., Revs. Mods. Phys. 53, 243 (1981);
d) de Vega, H.J., "Integrable QFT and Statistical Models", LPTHE preprint 85/54 (1985).

2) Baxter, R.J., "Exactly Solvable Models in Statistical Mechanics", Academic Press (1982).

3) Izergin, A.G. and Korepin, V.E., Comm. Math. Phys. 94, 67 and 93 (1984) and 99, 271 (1985).

4) Blöte, H.W.J., Cardy, J.L. and Nightingale, M.P., Phys. Rev. Lett. 56, 742 (1986);
Cardy, J.L., Nucl. Phys. B270, 186 (1986).

5) de Vega, H.J. and Woynarovich, F., Nucl. Phys. B251, 439 (1985).

6) Martin, H. and de Vega, H.J., Phys. Rev. B32, 5959 (1985).

7) de Vega, H.J. and Karowski, M., in preparation.

8) Karowski, M., Lecture given at the Paris-Meudon Colloquium, September 1986, to be published in the Proceedings (World Scientific).

9) de Vega, H.J., in preparation.

10) den Nijs, M.P.M., J. Phys. A12, 1857 (1979) and Phys. Rev. B27, 1674 (1983);
Gervais, J.-L. and Neveu, A., Nucl. Phys. B257, 59 (1985).

11) de Vega, H.J., Nucl. Phys. B240, 495 (1984).

12) Faddeev, L.D. and Takhtadzhyan, L.A., Russian Math. Surveys 34, 11 (1979).

13) Sutherland, B., Phys. Rev. B12, 3795 (1975);
Babelon, O., de Vega, H.J. and Viallet, C.M., Nucl. Phys. B200, 266 (1982); see also Ref. 1d);
Kulish, P.P. and Reshetikhin, N., J. Phys. A16, 2591 (1983).

14) Babujian, H.M. and Tsvelick, A.M., Nucl. Phys. B265, 24 (1986).

15) Baxter, R.J., Studies in Appl. Math. L 51, (1971); see also Ref. 2).

16) de Vega, H.J. and Lopes, E., CERN preprint TH.4567/86 (1986).

17) Babelon, O., de Vega, H.J. and Viallet, C.M., Nucl. Phys. B220, 13 (1983).

18) Woynarovich, F. and Eckle, M., Berlin preprint FUB (1986).

19) Yang, C.N. and Yang, C.P., Phys. Rev. 150, 321 (1966).

20) Hamer, C.J., Canberra preprint (1985).

OPERATOR CONTENT OF TWO-DIMENSIONAL POTTS MODELS

Bernard Nienhuis

Medische en fysiologische fysica,
Fysisch Laboratorium
Postbus 80.000
3508 TA Utrecht
NETHERLANDS

ABSTRACT

The central charge for the critical and the tricritical point of the general q-state Potts model is computed from the four-point energy correlation function. Thus the critical exponents for this class of models can be compared with the list of scaling dimensions permitted by unitarity and conformal invariance. It turns out that for the two- and three-state models all possible scaling dimensions are actually realized in at least one operator, but not in all cases in a scalar operator.

CLASSIFICATION OF MODULAR INVARIANT PARTITION FUNCTIONS

J.-B. Zuber

Service de Physique Théorique,
CEN-Saclay, 91191 Gif-sur-Yvette Cedex, France

ABSTRACT

Two-dimensional minimal conformal invariant theories are classified, using the constraint of modular invariance of their partition function on a torus.

1. INTRODUCTION

Conformal field theory is an essential piece in the construction of string theory[1]. It is also of importance in the study of two-dimensional critical systems, since dilatation invariance implies conformal invariance[2].

A conformal invariant theory in the Euclidean plane contains fields that transform covariantly under independent conformal transformations of the variables $z = x^1 + ix^2$ and $\bar{z} = x^1 - ix^2$:

$$z \to z' = f(z) \qquad \text{and} \qquad \bar{z} \to \bar{z}' = g(\bar{z}) \tag{1}$$

In the quantum theory, these transformations are realized by two sets of Virasoro generators L_n and $\bar{L}_n$ satisfying

$$\begin{aligned} [L_n, L_m] &= (n-m)\, L_{n+m} + \frac{c}{12}\, n\left(n^2-1\right) \delta_{n+m,0} \\ [\bar{L}_n, \bar{L}_m] &= (n-m)\, \bar{L}_{n+m} + \frac{c}{12}\, n\left(n^2-1\right) \delta_{n+m,0} \\ [L_n, \bar{L}_m] &= 0 \end{aligned} \tag{2}$$

The c-number "central charge" c is the coefficient of an anomalous term, that arises in the quantum theory.

A conformal theory[3] in the complex plane is specified by giving
i) the value of c
ii) the finite or infinite set of "primary" fields which transform[4] as $(h,\bar{h})$-tensors under (1):

$$\varphi_{h,\bar{h}}(z,\bar{z})\, dz^{h}\, d\bar{z}^{\bar{h}} = \varphi'_{h,\bar{h}}(z',\bar{z}')\, dz'^{h}\, d\bar{z}'^{\bar{h}} \tag{3}$$

or of the corresponding highest weight states of the two Virasoro algebras $V \otimes \bar{V}$:

$$\left|h,\bar{h}\right\rangle = \text{"}\lim_{z,\bar{z}\to 0}\text{"}\ \varphi_{h,\bar{h}}\ |0\rangle$$

(see [5] for a detailed discussion of the limit involved).
iii) the operator algebra[3]

$$\varphi^{(a)}(z,\bar{z})\ \varphi^{(b)}(0) = \sum_{c} C_{abc}\ z^{\Delta_{abc}}\ \bar{z}^{\bar{\Delta}_{abc}}\ \varphi^{(c)}(0) \tag{4}$$

where Δ and $\bar{\Delta}$ are simple combinations of the dimensions of the (primary or non primary) fields $\varphi^{(a,b,c)}$. The "structure constants" C_{abc} determine this algebra.

What are the consistent conformal theories? Two approaches have been followed to tackle this question. The first ones makes use of the crossing symmetry constraint i.e. of the associativity of the operator algebra: the structure constants satisfy non-linear relations. Classes of solutions have been discovered in this way[6], but a general classification of solutions seems difficult. On the other hand, Cardy [7] has suggested to consider a finite geometry, namely to put the conformal theory in a box with periodic boundary conditions, i.e. on a torus, and to use the modular invariance of the partition function to constraint the operator content of the theory. This approach, that we developped with A. Cappelli, C. Itzykson and H. Saleur[8-10] is quite effective and seems to lead to an exhaustive classification, at least of the minimal conformal theories to be defined soon.

Let us examine how a conformal theory may be defined on a torus. The torus is defined by two of its complex periods ω_1 and ω_2, chosen such that $\tau = \omega_2/\omega_1$ lies in the upper half-plane. If we want to make use of the formalism developped for the complex plane, we have to open the torus by cutting it along one of its geodesics, say ω_1. The resulting piece of cylinder may then be mapped onto an annulus in the plane by the exponential mapping:

$$w \to z = \exp 2\pi i \ w/\omega_1 \tag{5}$$

In this mapping, the translation operator along ω_2, i.e. the "Hamiltonian", is mapped on a certain combination of dilatation and rotation operators in the plane. To make things precise, we have to remember[3] that the stress-energy "tensors"

$$T(z) = \sum_{-\infty}^{\infty} z^{-n-2} L_n \qquad \bar{T}(\bar{z}) = \sum_{-\infty}^{\infty} \bar{z}^{-n-2} \bar{L}_n \tag{6}$$

are *not* primary fields but transform under (1) as:

$$T(z) = T'(z')\left(\frac{dz'}{dz}\right)^2 + \frac{c}{12}\left\{\frac{d^3z'/dz^3}{dz'/dz} - \frac{3}{2}\left(\frac{d^2z'/dz^2}{dz'/dz}\right)^2\right\} \tag{7}$$

Applied to the mapping (5), this leads to the following relations between the Virasoro generators on the cylinder and in the plane

$$\begin{aligned} L_{-1}^{cyl} &= \frac{1}{2\pi i}\int_{\text{geodesic } \omega_1} T^{cyl}(w)\, dw \\ &= \frac{1}{\omega_1}\oint_0 \frac{dz}{z}\left(T^{pl}(z)\, z^2 - \frac{c}{24}\right) \\ &= \frac{2\pi i}{\omega_1}\left(L_0^{pl} - \frac{c}{24}\right) \end{aligned} \tag{8}$$

The translation operator along ω_2 is therefore

$$H = \omega_2\, L_{-1}^{cyl} + \bar{\omega}_2\, \bar{L}_{-1}^{cyl} \tag{9}$$

$$= 2\pi i\left[\tau\left(L_0^{pl} - \frac{c}{24}\right) + \bar{\tau}\left(\bar{L}_0^{pl} - \frac{c}{24}\right)\right]$$

It is now natural to define the partition function of the system on *the torus* by

$$\begin{aligned} Z &= \mathrm{Tr}\ e^{H} \\ &= \left(q\bar{q}\right)^{-c/24}\ \mathrm{Tr}\left(q^{L_0}\bar{q}^{\bar{L}_0}\right) \end{aligned} \tag{10}$$

where $q \equiv \exp 2\pi i\tau$. The expression (10) is very general but quite implicit, since all the content of the theory is hidden in the symbol "Tr": the trace must be taken in the Hilbert space of states, which depends on the theory. As this Hilbert space carries a representation of the two Virasoro algebras $V \otimes \bar{V}$, we decompose it into a sum of irreducible representations labelled by highest weights $\left(h,\bar{h}\right)$, of multiplicities $\mathcal{N}_{h\bar{h}}$ (non negative integers). Thus

$$Z = \sum_{h,\bar{h}} \mathcal{N}_{h\bar{h}}\ \chi_h(q)\ \chi_{\bar{h}}^{*}(q) \tag{11}$$

where

$$\chi_h(q) = q^{-c/24}\ \left(\mathrm{tr}\ q^{L_0}\right)_{\mathrm{irrep.}\{h\}} \tag{12}$$

The Virasoro characters (defined here with an unconventional prefactor $q^{-c/24}$) may be regarded as generating functions of the number of states in the irreducible representation of highest weight h:

$$\chi_L(q) = q^{h-c/24} \sum_{n=0}^{\infty} d_n\ q^n \tag{13}$$

Here d_n is the dimension of the space of eigenstates of L_0 of eigenvalue h+n.

I recall that explicit formulae exist for all the possible cases [10].

In addition to the positivity of its entries, $\mathcal{N}_{h\bar{h}}$ must be a symmetric matrix (reality of Z), and $\mathcal{N}_{00}$ must be equal to 1, to express the unicity of the "vacuum" $h = \bar{h} = 0$). (In non-unitary theories, there may be states with negative h's, and the interpretation of this condition may be questionable).

Finally, one notices that in the calculation of the partition function on the torus, the two periods have not been treated on the same footing: the periodicity along ω_2 has been enforced by the trace operation. The partition function Z, however, should not depend on this choice. More generally, Z should be attached to the lattice generated by ω_1 and ω_2 but not depend on the particular choice of the periods ω_1, ω_2. In other words Z should be invariant under unimodular changes of the periods:

$$\left.\begin{aligned} \omega_2' &= a\,\omega_2 + b\,\omega_1 \\ \omega_1' &= c\,\omega_2 + d\,\omega_1 \end{aligned}\right\} \quad \begin{aligned} a, b, c, d \in \mathbb{Z} \\ ad - bc = 1 \end{aligned} \tag{14}$$

Since Z is a function of the dimensionless ratio $\tau = \frac{\omega_2}{\omega_1}$, it should be invariant under modular transformations of τ:

$$\tau = \frac{a\tau+b}{c\tau+d} \tag{15}$$

These modular transformations form an infinite discrete group, the modular group, denoted Γ; this group is generated for example by the two transformations $\tau \to \tau+1$ and $\tau \to -1/\tau$. The fact that a function of the form (11) must be a modular invariant turns out to be a very strong constraint. We shall examine its consequences for the simplest conformal theories: the "minimal" theories[3].

2. MINIMAL CONFORMAL AND A_1^1 AFFINE THEORIES

Minimal conformal theories are those containing a finite number of *primary* fields. As shown by Cardy [7], the modular invariance of (11) forces c to be less than one. Belavin, Polyakov and Zamolodchikov have introduced the set of theories of central charge

$$c = 1 - \frac{6(p-p')^2}{pp'} \tag{16}$$

where p and p' are coprime integers, and shown that the operator algebra closes for the finite number of primary fields $\varphi_{h,\bar{h}}$, where h and $\bar{h}$ take their values in the "Kac'table"[11]:

$$h_{rs} = h_{p'-r,p-s} = \frac{(pr-p's)^2-(p-p')^2}{4pp'} \tag{17}$$

with the restriction

$$1 \leq r \leq p'-1 \qquad\qquad 1 \leq s \leq p-1 \tag{18}$$

It is likely that those are the only theories with a finite number of primary fields. An important subset of the minimal theories is made of the $c < 1$ unitary theories[12] for which p and p' must be consecutive integers: $|p-p'| = 1$.

It is convenient to replace the pair of labels (r,s) by a single variable λ defined modulo $N = 2pp'$

$$\lambda = rp - sp' \tag{19}$$

That r,s may be recovered from λ results from the existence of ω_0, also defined mod N, and satisfying

$$\begin{aligned} rp + sp' &= \omega_0 \lambda \quad \bmod N \\ \omega_0^2 &= 1 \quad\quad \bmod 2N \end{aligned} \tag{20}$$

This number is easily constructed; given the pair of coprimes p and p', there exist a and b such that

$$ap - bp' = 1$$

Then $\omega_0 = ap + bp'$ does the job.

We then introduce the set of functions:

$$K_\lambda(\tau) = \frac{1}{\eta(\tau)} \sum_{-\infty}^{\infty} \exp\left[2i\pi\tau(Nn+\lambda)^2/2N\right] \tag{21}$$

where η is Dedekind's function:

$$\eta(\tau) = q^{1/24} \prod_{1}^{\infty} (1-q^n), \qquad q = e^{2\pi i\tau} \tag{22}$$

In terms of these functions, the conformal characters read

$$\chi_\lambda(\tau) = K_\lambda(\tau) - K_{\omega_0\lambda}(\tau) \tag{23}$$

They satisfy periodicity and reflection properties:

$$\chi_\lambda(\tau) = \chi_{\lambda+N}(\tau) = \chi_{-\lambda}(\tau) = -\chi_{\omega_0\lambda}(\tau) \tag{24}$$

These properties enable us to restrict ultimately the values of λ to a fundamental domain (represented by (18)) but for the time being, we only use the periodicity in $\lambda \to \lambda+N$.

It is not difficult to see that these characters transform under the two generators $T: \tau \to \tau+1$ and $S: \tau \to -\frac{1}{\tau}$ of the modular group Γ according to

$$\begin{aligned} \chi_\lambda(\tau+1) &= \exp 2i\pi\left(\frac{\lambda^2}{2N} - \frac{1}{24}\right) \chi_\lambda(\tau) \\ \chi_\lambda(-1/\tau) &= \frac{1}{\sqrt{N}} \sum_{\lambda' \in Z/NZ} \exp 2i\pi\, \lambda\lambda'/N \qquad \chi_{\lambda'}(\tau) \end{aligned} \tag{25}$$

i.e. they form a unitary representation of Γ.

There is an unexpected connection between the construction of conformal modular invariants and that of modular invariants made of A_1^1 Kac-Moody characters. In their study of the string compactification on a SU(2) group manifold, Gepner and Witten[13] have shown that the one-loop partition function of the group degrees of freedom takes the form

$$Z = \sum \mathcal{N}_{\lambda\bar{\lambda}} \, \chi_\lambda(\tau) \; \chi^*_{\bar{\lambda}}(\tau) \tag{26}$$

Here $\lambda = 2l+1$, where l is the integer or half-integer spin labelling the integrable representations of $A_1^{(1)}$ of central charge k ($\lambda \le k+1$); χ_λ are the affine characters[14], up to a certain power of q; $\mathcal{N}_{\lambda\bar{\lambda}}$ are non negative integers, with again $\mathcal{N}_{11} = 1$, and Z must be modular invariant. As we shall see there is a direct relation between the conformal and the affine invariants[15]. That there are connections between the minimal conformal and A_1^1 theories was actually already noticed or used in Ref.[10c,16]. At any rate, it is natural to treat both problems in parallel, and to introduce convenient notations for this purpose. In the A_1^1 case, we denote

$$N = 2(k+2)$$

The variable λ takes its values in a fundamental domain

$$1 \le \lambda \le k+1 \tag{27}$$

but it is actually suitable to extend it to a variable of $\mathbb{Z}/N\mathbb{Z}$. The characters read[13-15]

$$\chi_\lambda(\tau) = \frac{1}{\eta^3(\tau)} \sum_{-\infty}^{\infty} (nN+\lambda) \; e^{2i\pi(nN+\lambda)^2/N} \tag{28}$$

They satisfy

$$\chi_\lambda = \chi_{\lambda+N} = -\chi_{-\lambda} \tag{29}$$

and they have the following modular transformations

$$\begin{aligned} \chi_\lambda(\tau+1) &= \exp 2i\pi\left(\frac{\lambda^2}{2N} - \frac{1}{8}\right) \chi_\lambda(\tau) \\ \chi_\lambda\left(-\frac{1}{\tau}\right) &= \frac{-i}{\sqrt{N}} \sum_{\lambda'\in\mathbb{Z}/N\mathbb{Z}} \exp 2i\pi\, \lambda\lambda'/N \;\; \chi_{\lambda'}(\tau) \end{aligned} \tag{30}$$

and form also a finite dimensional unitary representation of Γ.

In both the conformal and the affine cases, one may show that the infinite subgroup Γ_{2N} of Γ, made of the transformations (15) with

$$a = d = \pm 1 \mod 2N, \qquad b = c = 0 \mod 2N$$

changes the characters only by a τ- and λ-independent phase. This is particularly clear for the transformation T^{2N} for which a=d=1, b=2N, c=0 and $\chi_\lambda(\tau+2N) = \exp -2i\pi\, N/12\ \ \chi_\lambda(\tau)$. The subgroup Γ_{2N}, however, is *not* generated by T^{2N} and its conjugates, for $2N \geq 6$, and this property requires therefore an explicit proof[9]. It implies that only the finite coset group

$$\Gamma/\Gamma_{2N} = PSL\ (2,\ \mathbb{Z}/2N\mathbb{Z})$$

acts non trivially on characters through a projective unitary representation U. Modular invariance of (11) or (26) is thus equivalent to

$$U_{\lambda\lambda'}(A)\ \mathcal{N}_{\lambda'\bar{\lambda}'}\ U^{\dagger}_{\bar{\lambda}'\bar{\lambda}}(A) = \mathcal{N}_{\lambda\bar{\lambda}}$$

or (31)

$$U(A)\mathcal{N} = \mathcal{N}\ U(A)$$

for any $A \in \Gamma/\Gamma_{2N}$. The forthcoming discussion is just tantamount to a study of the commutant of this projective representation.

3. GENERAL FORM OF THE INVARIANTS

Let us first consider the implications of the invariance under T: $T\mathcal{N}T^{\dagger} = \mathcal{N}$. From Eqs.(25) or (30), we see that $\mathcal{N}_{\lambda\bar{\lambda}} \neq 0$ only if

$$\lambda^2 = \bar{\lambda}^2 \mod 2N \tag{32}$$

It is not difficult[9] to show that this condition is equivalent to the following set of conditions:

i) there exists an integer α, divisor of λ and $\bar{\lambda}$, such that α^2 divides $\frac{N}{2}$

ii) there exists μ defined modulo N/α^2, such that $\mu^2 = 1 \mod 2N/\alpha^2$

iii) $$\frac{\bar{\lambda}}{\alpha} = \mu \frac{\lambda}{\alpha} \mod \frac{N}{\alpha^2}$$

or equivalently, there exists ξ defined mod α, such that

$$\bar{\lambda} = \mu \lambda + \xi \frac{N}{\alpha} \mod N$$

In other words, invariance under T forces $\mathcal{N}$ to have the general form:

$$\mathcal{N}_{\lambda\bar{\lambda}} = \sum_{\alpha,\mu,\xi} C_{\alpha,\mu,\xi,\lambda} \, \delta_{\bar{\lambda},\mu\lambda+\xi\frac{N}{\alpha}} \tag{33}$$

with α,μ,ξ satisfying the conditions above.

It seems that invariance under S forces $C_{\alpha,\mu,\xi,\lambda}$ in (33) to be independent of ξ and λ. This has been proved in the case where $\frac{N}{2}$ has no square divisor ($\alpha = 1$ is the only possibility) but is likely to be general. We thus conjecture that the general solution is

$$\mathcal{N}_{\lambda\bar{\lambda}} = \sum_{\substack{\alpha,\mu \\ \text{as above}}} C_{\alpha,\mu} \sum_{\xi\in\mathbb{Z}/\alpha\mathbb{Z}} \delta_{\bar{\lambda},\mu\lambda+\xi\frac{N}{\alpha}} \tag{34}$$

This conjecture has two immediate consequences:

i) If

$$N = \prod_{i=1}^{n} p_i^{r_i} \tag{35}$$

is the decomposition of N into a product of powers of primes p_i, the number of linearly independent invariants in the A_1^1 case is simply

$$\psi\left(\frac{N}{2}\right) = \frac{1}{2}\left\{\prod_{i=1}^{n} (1+r_i)-\delta\right\} \tag{36}$$

with $\delta = 1$ if $\frac{N}{2}$ is a square, 0 otherwise. For large N, on the average,

$\phi(N/2)$ grows as $\frac{1}{2}\ ln\ N$.

ii) For the conformal invariants, one has first to reexpress the invariant associated with (34) in terms of independent characters satisfying the condition (18). It turns out that the matrix $\mathcal{N}$ then factorizes as:

$$\mathcal{N}^{conf}_{rs,r's'} = \sum \mathcal{N}_{rr'}\left(\frac{N}{2} = p'\right) \cdot \mathcal{N}_{ss'}\left(\frac{N}{2} = p\right) \tag{37}$$

i.e. as a sum of tensor products of invariants of the form (34) relative to the indices $r,\bar{r}$ and $s,\bar{s}$. This is the remarkable connection between affine and conformal invariants alluded to above.

The number of independent conformal invariants is thus $\phi(p')\ \phi(p)$.

4. POSITIVE MODULAR INVARIANTS

We now enforce the two constraints $\mathcal{N}_{\lambda\bar{\lambda}} \geq 0$ for λ, $\bar{\lambda}$ in the fundamental domain (18) or (27) and $\mathcal{N}_{00} = 1$. The invariants listed in Table 1 and 2 satisfy these constraints. We conjecture that these lists are exhaustive! We are led to this conjecture by:

i) the inspection of all affine invariants up to $\frac{N}{2} = 100$

ii) The observation of a totally unexpected connection with the classification of simply laced Lie algebras. Not only our affine solutions come in two infinite series plus three exceptional cases, but the values of the indices $\lambda = \bar{\lambda}$ carried by the *diagonal* terms in Table 1 are the exponents of the Lie algebras A, D, E[17]. For an algebra of rank r, there are r exponents e_i, such that $1 + e_i$ give the degrees of a set of independent generators of invariant polynomials (or "higher Casimir invariants"). In particular, the algebra $D_{2\rho+2}$ has two independent invariants of degree $(2\rho+1)+1$, and this agrees nicely with the factor 2 in the corresponding invariant of Table 1.

The affine invariants associated with the algebra A are nothing but the diagonal sum of all $|\chi_\lambda|^2$. Other solutions exist only for even $\frac{N}{2}$. The conformal invariants are then obtained by tensor products of affine invariants pertaining to the values p' and p of $k+2 = \frac{N}{2}$. As p and p' are coprimes, they cannot be simultaneously even, and conformal invariants are thus labelled by pairs of simply laced algebras (A,A),

(A,D), (A,E),... with at least one A entry. (see Table 2).

To summarize, we believe we now have a full classification of all minimal conformal theories, i.e. of all universality classes of critical theories with a finite number of primary fields.

What do these new theories describe?

In the affine case, the A and D series were interpreted[10] as the Wess-Zumino theory on SU(2) and SO(3) respectively, while the E_6 and E_8 cases are associated with embeddings of SU(2) into SU(3) and G_2 [18]; the case E_7 remains unexplained.

In the conformal case, and for unitary theories (p' = m, p = m+1 or vice versa), the (A,A) series describes the critical Ising model or its RSOS generalizations[19]. The (A_4,D_4) and (D_4,A_6) theories represent the critical and tricritical 3-state Potts models. What about the others? In recent papers[20], Pasquier has introduced a new family of completely integrable lattice models, in which Dynkin diagrams of simply laced algebras encode the selection rules between states at neighbouring sites. He suggested that the critical behaviour of these theories is described by the unitary conformal theories just described. Computation of critical exponents in these models would confirm this suggestion.

Finally, I want to return to the yet mysterious classification of invariants in terms of simply laced Lie algebras. It does not mean that the corresponding algebra is a symmetry of the theory but rather gives one more example that the (A,D,E) classification appears in a great variety of apparently unrelated problems[21].

We have seen how powerful the constraint of modular invariance is, in the simplest case of minimal conformal theories. It seems natural to apply it to superconformal[22] and to non minimal theories[23].

ACKNOWLEDGEMENTS

The work reported here has been made in a very enjoyable collaboration with A. Cappelli and C. Itzykson. I have also benefited from discussions or communications with D. Bernard, V. Dobrev, V. Kac, W. Nahm, V. Pasquier, H. Saleur and A. Schwimmer, whom I am glad to thank here.

REFERENCES

[1] **Shenker S., Contribution to this meeting**
[2] **Polyakov A.M., Zh. E.T.F. Lett.** 12, 538 (1970)
(J.E.T.P. Lett. 12, 381 (1970)
[3] Belavin A.A., Polyakov A.M. and Zamolodchikov A.B., Nucl. Phys. B241, 333 (1984)
[4] Gervais J.L. and Sakita B., Nucl. Phys. B34, 477 (1971)
[5] Nahm W., Contribution to this meeting
[6] Dotsenko V.S., Nucl. Phys. B235 [FS11], 54 (1984)
Dotsenko V.S. and Fateev V.A., Nucl. Phys. B240 [FS12],312 (1984), B251 [FS13] 691 (1985)
Fateev V.A. and Zamolodchikov A.B., Zh. ET.F. 89, 380 (1985), (JETP. 62, 215 (1985)) and Landau Institute preprint
[7] Cardy J.L., Nucl. Phys. B270 [FS16] 186 (1986)
[8] Itzykson C. and Zuber J.-B., Nucl. Phys. B275 [FS17] 580 (1986)
Itzykson C., Saleur H. and Zuber J.-B., Europhysics Lett. 2, 91 (1986)
[9] Cappelli A., Itzykson C., and Zuber J.-B., Nucl. Phys. B, to appear
[10] a. Feigin V.G. and Fuchs D.B., Functs. Anal. Prilozhen 17, 91 (1983) (Funct. Anal. and Appl. 17, 241 (1983)
b. Rocha-Caridi A., in "Vertex Operators in Mathematics and Physics" edited by Lepowsky J., Mandelstam S. and Singer I., Springer Verlag, NY, 1985; and contribution to this meeting
c. Dobrev V., in Proceedings of the XIII International Conference on Differential Geometric Methods in Theoretical Physics (Shumen 1984), edrs. H.D. Doebner and P.D. Palev, World Scientific, Singapour 1986
[11] Kac V., Lecture Notes in Physics 94, 441 (1979)
Feigin B.L. and Fuchs D.B., Funct. Anal. Priloshen 16, 47 (1982) (Funct. Anal. and Appl. 16, 114 (1982))
[12] Friedan D., Qiu Z. and Shenker S., Phys. Rev. Lett. 52, 1575 (1984) and in "Vertex Operators in Mathematics and Physics", op. cit.
[13] Gepner G. and Witten E., Princeton preprint
[14] Kac V.G. and Peterson D., Adv. in Math. 53, 125 (1984)
Kac V.G., "Infinite Dimensional Lie Algebras", Cambridge University Press, London 1985
[15] Gepner G., Princeton preprint
[16] Goddard P., Kent A., and Olive D., Phys. Lett. 152B, 88 (1985), Comm. Math. Phys. 103, 105 (1986)
[17] Kac V., (private communication) made first this observation in

the m=11 conformal theory
[18] Nahm W. and Bernard D., private communications
[19] Andrews G.E., Baxter R.J. and Forrester P.J., J. Stat. Phys. 35, 193 (1984)
Huse D.A., Phys. Rev. B30, 3908 (1984)
[20] Pasquier V., Saclay preprints PhT/86-124 and 86/139
[21] Arnold V.I., "Catastrophe Theory", Springer Verlag, Berlin 1984
[22] Cappelli A., Saclay preprint PhT/86-149
[23] Di Francesco P. and Zuber J.-B., to appear

TABLE 1

List of known partition functions in terms of $A_1^{(1)}$ characters

$k \geq 1$	$\sum_{\lambda=1}^{k+1} \lvert\chi_\lambda\rvert^2$	A_{k+1}
$k = 4\rho,\ \rho \geq 1$	$\sum_{\substack{\lambda\ \mathrm{odd}\, =\, 1 \\ \lambda \neq 2\rho+1}}^{4\rho+1} \lvert\chi_\lambda\rvert^2 + 2\lvert\chi_{2\rho+1}\rvert^2 + \sum_{\lambda\ \mathrm{odd}\, =\, 1}^{2\rho-1} (\chi_\lambda \chi^*_{4\rho+2-\lambda} + \mathrm{c.c.})$ $= \sum_{\lambda\ \mathrm{odd}\, =\, 1}^{2\rho-1} \lvert\chi_\lambda + \chi_{4\rho+2-\lambda}\rvert^2 + 2\lvert\chi_{2\rho+1}\rvert^2$	$D_{2\rho+2}$
$k = 4\rho-2,\ \rho \geq 2$	$\sum_{\lambda\ \mathrm{odd}\, =\, 1}^{4\rho-1} \lvert\chi_\lambda\rvert^2 + \lvert\chi_{2\rho}\rvert^2 + \sum_{\lambda\ \mathrm{even}\, =\, 2}^{2\rho-2} (\chi_\lambda \chi^*_{4\rho-\lambda} + \mathrm{c.c.})$	$D_{2\rho+1}$
$k+2 = 12$	$\lvert\chi_1+\chi_7\rvert^2 + \lvert\chi_4+\chi_8\rvert^2 + \lvert\chi_5+\chi_{11}\rvert^2$	E_6
$k+2 = 18$	$\lvert\chi_1+\chi_{17}\rvert^2 + \lvert\chi_5+\chi_{13}\rvert^2 + \lvert\chi_7+\chi_{11}\rvert^2 + \lvert\chi_9\rvert^2 + [(\chi_3+\chi_{15})\chi_9^* + \mathrm{c.c.}]$	E_7
$k+2 = 30$	$\lvert\chi_1+\chi_{11}+\chi_{19}+\chi_{29}\rvert^2 + \lvert\chi_7+\chi_{13}+\chi_{17}+\chi_{23}\rvert^2$	E_8

TABLE 2

List of known partition functions in terms of conformal characters. The unitary series corresponds to p' = m+1, p = m or p = m+1, p' = m, m = 3,4,....

Condition	Partition function	Type
	$\frac{1}{2}\sum_{r=1}^{p'-1}\sum_{s=1}^{p-1} \lvert\chi_{rs}\rvert^2$	$(A_{p'-1}, A_{p-1})$
$p' = 4\rho+2$, $\rho\geq 1$	$\frac{1}{2}\sum_{s=1}^{p-1}\left\{\sum_{r\ \mathrm{odd}=1,\ r\neq 2\rho+1}^{4\rho+1} \lvert\chi_{rs}\rvert^2 + 2\lvert\chi_{2\rho+1\ s}\rvert^2 + \sum_{r\ \mathrm{odd}=1}^{2\rho-1}(\chi_{rs}\chi^*_{p'-r\ s}+\mathrm{c.c.})\right\}$	$(D_{2\rho+2}, A_{p-1})$
$p' = 4\rho$, $\rho\geq 2$	$\frac{1}{2}\sum_{s=1}^{p-1}\left\{\sum_{r\ \mathrm{odd}=1}^{4\rho-1} \lvert\chi_{rs}\rvert^2 + \lvert\chi_{2\rho\ s}\rvert^2 + \sum_{r\ \mathrm{even}=1}^{2\rho-2}(\chi_{rs}\chi^*_{p'-r\ s}+\mathrm{c.c.})\right\}$	$(D_{2\rho+1}, A_{p-1})$
$p' = 12$	$\frac{1}{2}\sum_{s=1}^{p-1}\left\{\lvert\chi_{1s}+\chi_{7s}\rvert^2 + \lvert\chi_{4s}+\chi_{8s}\rvert^2 + \lvert\chi_{5s}+\chi_{11s}\rvert^2\right\}$	(E_6, A_{p-1})
$p' = 18$	$\frac{1}{2}\sum_{s=1}^{p-1}\left\{\lvert\chi_{1s}+\chi_{17s}\rvert^2 + \lvert\chi_{5s}+\chi_{13s}\rvert^2 + \lvert\chi_{7s}+\chi_{11s}\rvert^2 + \lvert\chi_{9s}\rvert^2 + [(\chi_{3s}+\chi_{15s})\chi^*_{9s}+\mathrm{c.c.}]\right\}$	(E_7, A_{p-1})
$p' = 30$	$\frac{1}{2}\sum_{s=1}^{p-1}\left\{\lvert\chi_{1s}+\chi_{11s}+\chi_{19s}+\chi_{29s}\rvert^2 + \lvert\chi_{7s}+\chi_{13s}+\chi_{17s}+\chi_{23s}\rvert^2\right\}$	(E_8, A_{p-1})

STRING MODELS WITH TWISTED VERTEX OPERATORS*

D. Altschüler, Ph. Béran, J. Lacki, I. Roditi**

Département de Physique Théorique
Université de Genève
1211 Genève 4, Switzerland.

ABSTRACT

We show how to derive in a systematic way all possible symmetry breaking patterns of the heterotic string, which arise from twisted boundary conditions on the torus. Our method uses twisted Vertex Operator constructions of affine Lie algebras, which are generalizations of the Frenkel-Kac construction.

---

* Talk presented by the first author.

** permanent address : Centro Brasileiro de Pesquisas Fisicas, CBPF, Rio de Janeiro, 22290, Brazil.

1. INTRODUCTION.

At present, it seems that the heterotic string is the favorite candidate for a unified theory. This model has very rich algebraic and geometrical structures built in, and it would be nice to reach a better understanding of these. One particularly important problem is the fact that the gauge groups $E_8 \times E_8$ or O(32) are too large, so that we need some kind of symmetry breaking process to occur which could bring us closer to the "low-energy" world. It is also natural to ask for a description of this process within string theory, not after taking a field theory limit.

Looking back from the late sixties onwards, one sees frequent interactions between string theory and mathematics. An important concept used in both areas, though for different reasons, is the Vertex Operator. Physicists discovered it as a mean of describing interactions, and it became a tool for mathematicians studying Kac-Moody algebras. The Frenkel-Kac Vertex Operator construction[1)] is popular among physicists since it is a key ingredient of the heterotic string[2)]. Another construction[3)] involving oscillators with half integral modes was also applied to physics, e.g. in ref. 4,5. The most general Vertex construction of affine Lie algebras has recently been worked out[6,7)], so that all previously known constructions now can be regarded as particular cases. In this paper we will show that this general construction is the appropriate mathematical setting to discuss the possible symmetry breaking patterns of the heterotic string, or more generally, of any theory part of whose degrees of freedom are strings moving on a torus.

Let us briefly review the salient features of the motion of a closed string on a torus and its quantization, the Frenkel-Kac construction. In what follows, d shall denote the dimension of the torus T^d :

$$T^d = \mathbf{R}^d \; / \; \Gamma \tag{1}$$

where Γ is a lattice. In order to obtain after quantization states belonging to representations of a semisimple Lie group G, we take Γ to be the weight lattice of G. The Lie algebra of G we write g. The coordinates of the string in $\mathbf{R}^d$ have the mode expansion

$$X_I(\sigma,\tau) = x_I + p_I\tau + 2L_I\sigma + oscillators \qquad I=1,\ldots,d \quad (2)$$

where $L \in \Gamma$, so that the string is closed on T^d. Also we see that $p \in \Gamma^* = Q$ the root lattice of G. In the heterotic string one keeps only the left-movers :

$$X_I(\sigma,\tau) = x_I + p_I(\tau+\sigma) + left-oscillators \qquad (2')$$

The Hilbert space has the following structure :

$$H = F \otimes \mathrm{C}(Q). \qquad (3)$$

F is the Fock space representation of the Heisenberg algebra spanned by the oscillators and the identity operator. $\mathrm{C}(Q)$ is an infinite-dimensional vector space spanned by vectors $|k>$, $k \in Q$ which are momentum eigenstates :

$$p_I\,|k> = k_I\,|k>. \qquad (4)$$

In other words,

$$H = \bigoplus_{k \in Q} F_k \qquad (5)$$

where each Fock space F_k has vacuum $|k>$, and is a copy of F. This latter fact is a source of trouble if you want to construct non-simply laced affine algebras. See D. Olive's lecture in these proceedings and ref. 8 for more about this. From now on, we assume g is simply-laced.

The translations group Q is represented by the operators :

$$\exp(ik^I x_I)\,|k'> = |k+k'> \qquad k,k' \in Q. \qquad (6)$$

The Vertex Operator is in this case :

$$V(k,z) = :\exp\left[-\sum_{n\neq 0} \frac{z^{-n}}{n} k(n) \right] : \otimes T_k(z) \tag{7}$$

k(n) is an abbreviation for $\sum_I k_I \alpha_n^I$, where $k \in Q$ and α_n^I are the oscillators. The factor $T_k(z)$ acts in $\mathbf{C}(Q)$ only and is

$$T_k(z) = \exp[i\sum_I k_I (x_I - ip_I \log z)]\ c_k \tag{8}$$

c_k is the most subtle part of the construction. It has received many different names in the physics literature, e.g. Klein factor, cocycle or twist. It is defined as follows :

$$c_k\ |k'\rangle = \epsilon(k,k')\ |k'\rangle \tag{9}$$

ϵ is a map : $Q \times Q \to \{\pm 1\}$ satisfying :

$$\epsilon(k_1,k_2)\ \epsilon(k_1+k_2,k_3) = \epsilon(k_1,k_2+k_3)\ \epsilon(k_2,k_3) \tag{10a}$$

$$\epsilon(k,0) = 1 \tag{10b}$$

$$\epsilon(k,-k) = 1 \tag{10c}$$

$$\epsilon(k_1,k_2)\ /\ \epsilon(k_2,k_1) = (-1)^{(k_1|k_2)} \tag{10d}$$

$(\ |\ )$ is the scalar product in the euclidean space $\mathbf{R}^d$. Eqs. (10a,b) say that ϵ is a two-cocycle of the group Q. (10c) is a normalizing condition. (10d) uniquely characterizes the particular cohomology class of two-cocycles which is needed in this construction. We shall encounter other two-cocycles in this paper, which will share with ϵ properties (10a-c), but with the RHS of (10d) replaced by a bimultiplicative function $S(k_1,k_2)$ whose value is a complex phase, with $S(k,k) = 1$, i.e. S is alternating. Such "symmetry factors" and their relation with cocycles are discussed e.g. in the work of Goddard, Olive and Schwimmer[8]. S is also called "commutator map" in ref 7.

The reason ϵ and c_k are introduced is that in a Chevalley basis of g the structure constants are given by ϵ. More precisely, if Δ is the set of roots, and $\alpha,\beta \in \Delta$,

$$[E_\alpha, E_\beta] = \begin{cases} \epsilon(\alpha,\beta)E_{\alpha+\beta} & \text{if } \alpha+\beta \in \Delta \\ h_\alpha & \text{if } \alpha+\beta=0 \\ 0 & \text{if } \alpha+\beta \notin \Delta \end{cases} \tag{11}$$

The moments of the Vertex Operator $V(\alpha,z)$ $\alpha \in \Delta$,

$$E_\alpha(n) = \frac{1}{2\pi i} \oint dz\; z^{n-1}\, V(\alpha,z) \tag{12}$$

for $n \in \mathbf{Z}$, together with the oscillators and the identity form a representation of the affine algebra $\hat{g}$ constructed on g, which is equivalent to the basic representation of level one.

Notice that the operators :

$$\hat{c}_k = \exp(i\sum_I k_I x_I)\; c_k \tag{13}$$

form a projective representation of Q on $\mathbf{C}(Q)$,

$$\hat{c}_k\, \hat{c}_l = \epsilon(k,l)\; \hat{c}_{k+l} \tag{14}$$

Before coming to the generalization of the Vertex construction, it is useful to have in mind a very simple picture of H, obtained as follows. Recall the Virasoro algebra with generators :

$$L_n = \frac{1}{2} \sum_{m \in \mathbf{Z}} :\alpha^I_{n-m}\; \alpha^I_m: \tag{15}$$

Up to an additive constant, L_0 is the $(mass)^2$ operator,

$$L_0 = \frac{1}{2}p_I^{\,2} + \sum_{m=1}^{\infty} \alpha^I_{-m}\; \alpha^I_m \tag{16}$$

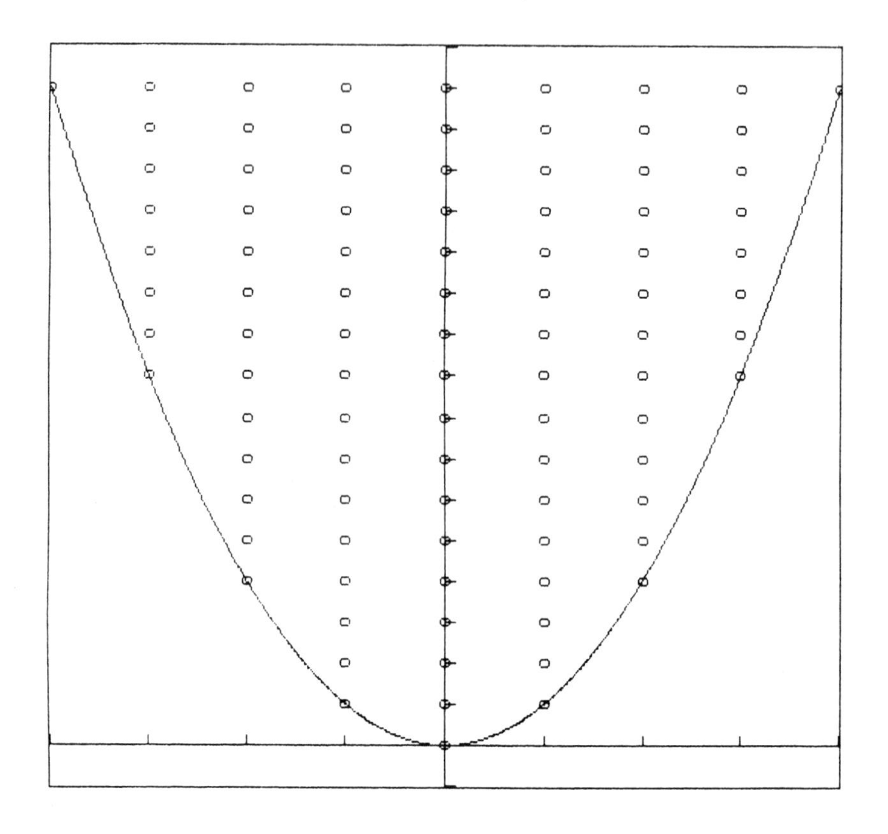

fig. 1 : The Hilbert space.

Decompose H into eigenstates of L_0 :

$$H = \bigoplus_{N=0}^{\infty} H_N \tag{17}$$

where H_N consists of states with eigenvalue N. Working out the commutation relations of L_n with $\hat{g}$, one finds that

$$[L_0, g] = 0 \tag{18}$$

g being represented by p_I and $E_\alpha(0)$. This means that H_N is invariant under g, i.e. it is a multiplet of states which transforms according to some representation of g. Therefore one can draw a picture of H as in fig. 1, where for simplicity we took Q to be one-dimensional. The horizontal axis is Q, the vertical axis the eigenvalues N. An open dot sitting at (k,N) where $k \in Q$ stands for all states in H with eigenvalues k and N of p and L_0. In particular, the open dots on the limiting paraboloid are the vacua $|k>$, and F_k consists of all dots above one such vacuum. Finally, we saw that the intersection of the paraboloid with horizontal planes at height N is a representation of g.

The partition function, which is called q-dimension or Poincaré series in mathematics is defined by

$$dim_q H = \sum_{N=0}^{\infty} (dim\ H_N)\ q^N \tag{19}$$

and given in this case by

$$dim_q H = \frac{\sum_{k \in Q} q^{\frac{1}{2}|k|^2}}{\prod_{j=1}^{\infty} (1 - q^j)^d} \tag{20}$$

The numerator is the theta function associated with the lattice Q and the denominator the contribution of the Fock space F.

2. TWISTED STRINGS.

In this section, we study the geometrical objects which are naturally associated to the general Vertex constructions: strings on the torus obeying twisted boundary conditions. Recall the torus (1), and denote by π the canonical projection :

$$\pi : \mathbf{R}^d \to T^d \tag{21}$$

$$x \longmapsto \tilde{x} = x \bmod \Gamma$$

Consider the group $Aut(\Gamma)$ of invertible linear maps $\Gamma \to \Gamma$ (automorphisms). Fix an element $w \in Aut(\Gamma)$ of order m, i.e. m is the least positive integer such that $w^m = 1$. Then w defines an automorphism $\tilde{w}$ of the torus T^d by :

$$\tilde{w}(\tilde{x}) = \pi(w(x)) \text{ for } x \in \pi^{-1}(\tilde{x}). \tag{22}$$

A twisted string by definition obeys the boundary conditions :

$$X(\pi,\tau) = \tilde{w} \cdot X(0,\tau), \tag{23}$$

i.e. it is closed up to the action of $\tilde{w}$. The same kind of boundary conditions has been studied before[5,9].

Since w preserves $(\mid)$ it is diagonalizable, its eigenvalues belong to the m-th roots of unity ω^j, $j=0,...,m-1$, $\omega = \exp(2\pi i/m)$. Let $\underline{h}$ be the complexification of $\mathbf{R}^d$. Decompose $\underline{h}$ into eigenstates :

$$\underline{h} = \bigoplus_{j=0}^{m-1} \underline{h}_j \tag{24}$$

where $\underline{h}_j$ is the eigenspace belonging to ω^j, which can be trivial for some values of j. Because w is an orthogonal transformation,

$$(\underline{h}_j \mid \underline{h}_k) \begin{cases} =0 & \text{if } j+k \neq 0 \; mod \; m \\ \neq 0 & \text{if } j+k = 0 \; mod \; m \end{cases} \tag{25}$$

$\underline{h}_j$ and $\underline{h}_{m-j}$ are actually dual to each other. For a vector $x \in \underline{h}$ we write x_j to denote its orthogonal projection on $\underline{h}_j$. For example x_0 is the projection on the invariant subspace $\underline{h}_0$, so that $w(x_0)=x_0$. Also we set $d_j = dim(\underline{h}_j)$.

It is clear that $X_0(\sigma,\tau)$ is an ordinary closed string, moving on a smaller torus of dimension d_0. Hence the mode expansion for these components (if they exist) is :

$$X_0(\sigma,\tau) = x_0 + p_0\tau + 2L_0\sigma + oscillators \tag{26}$$

where

$$L_0 \in \Gamma \cap \underline{h}_0 \equiv \Gamma^w, \tag{27}$$

and $x_0 \in \underline{h}_0$. Thus the momenta p_0 must lie on the dual of Γ^w *in* $\underline{h}_0$, which is nothing but Q_0, the projection of the root lattice Q on $\underline{h}_0$. In a heterotic theory one then writes an equ. (26') similar to (2').

The solution to the equations of motion in the twisted subspaces $\underline{h}_j$ *with* $j \neq 0$ is quite different. First, there is no winding vector L since the string is not closed in these sectors. Second, the center of mass position

$$x_f = \sum_{j=1}^{m-1} x_j \tag{28}$$

is "frozen". It must sit at some isolated fixed point of the transformation $\tilde{w}$ on the torus : $\tilde{w}(\tilde{x}_f) = \tilde{x}_f$. If $T^{\tilde{w}}$ is the set of fixed points, then

$$\tilde{x}_f \in T^{\tilde{w}} / \pi(\underline{h}_0). \tag{29}$$

The mode expansion (for left-movers) is

$$X_j(\sigma,\tau) = x_j + \frac{i}{2} \sum_{n \in \mathbf{Z}+j/m} \frac{1}{n} \alpha_n e^{-2in(\tau+\sigma)} \tag{30}$$

Notice the fractional moding by j/m of the oscillators, due to (23).

At this point, it is perhaps a good idea to work out the set of fixed points for some examples.

Example 1.

Take $w = r_\gamma$, the reflection in the hyperplane perpendicular to some $\gamma \in \Gamma$, $|\gamma|^2=2$:

$$r_\gamma(x) = x - (x|\gamma)\gamma \ , \ x \in \underline{h}.$$

Then w is of order two, $\underline{h}_0$ is the hyperplane perpendicular to γ, $d_0=d-1$, $\underline{h}_1$ is the line through γ, $d_1=1$ and since

$$x - w(x) = (x|\gamma)\gamma$$

one finds

$$T^{\tilde{w}} = \{\tilde{x} \in T^d \mid (x|\gamma) \in \mathbf{Z}, x = \pi^{-1}(\tilde{x})\}$$

Example 2.

w= -1 i.e. $w : x \mapsto -x$.

In this case, $\underline{h}_0 = 0$, $\underline{h}_1 = \underline{h}$. If $x - w(x) = 2x \in \Gamma$ then $x \in \frac{1}{2}\Gamma$. Hence,

$$T^{\tilde{w}} = \Gamma/2\Gamma$$

contains 2^d points.

It is useful to define the set :

$$M_w = \{x \in \underline{h} \mid x - w(x) \in \Gamma\} \tag{31}$$

Note that $M_w = \pi^{-1}(T^{\tilde{w}})$, i.e. when projected on the torus, this set gives the fixed points. We shall regard it as an abelian group.

3. QUANTIZATION.

From this point we assume that the lattice Q is even and sel-dual ($Q = \Gamma$), the latter assumption being made to avoid unnecessary technical complications. The Hilbert space has the general form analogous to (5) :

$$H_w = \bigoplus_{cm} F_{cm} \tag{32}$$

F_{cm} is a Fock space generated by the creation operators acting on a vacuum $|cm\rangle$, where cm stands for all center of mass quantum numbers, which we have to describe now. Actually, for all cm, F_{cm} is isomorphic to a generic Fock space F, so that (compare with (3)) :

$$H_w = F \otimes V, \tag{33}$$

and V is the space spanned by the $|cm\rangle$ states.

From the discussion in the previous section, it is obvious that in the center of mass position :

$$\tilde{x} = \tilde{x}_0 + \tilde{x}_f \tag{34}$$

one can choose independently $\tilde{x}_0$ and $\tilde{x}_f$. The first one specifies the cm position of the closed string X_0, so that the corresponding quantum state will be just $|k_0\rangle$, where $k_0 \in Q_0$ (remember (27)), and

$$p_{0,I} |k_0\rangle = k_{0,I} |k_0\rangle \tag{35}$$

where the subscript I refers to a component of the subspace $\mathfrak{h}_0$. The naive guess for the quantum state corresponding to the second term in (34) would be to take $|\tilde{x}_f>$, where $\tilde{x}_f$ is chosen in the discrete set (29). However, if we want to end up with H_w being the representation space of an affine algebra, it must be slightly modified as we now explain*.

Define a map

$$\psi : M_w \times M_w \to \mathbf{C}^*$$

$$\psi(x,y) = \exp 2\pi i(x|y - w(y)) \tag{36}$$

It has the property of being bimultiplicative,

$$\psi(x_1+x_2,x_3) = \psi(x_1,x_3)\ \psi(x_2,x_3) \tag{37}$$

$$\psi(x_1,x_2+x_3) = \psi(x_1,x_2)\ \psi(x_1,x_3)$$

and alternating,

$$\psi(x,x) = 1 \text{ for } any\ x \in M_w . \tag{38}$$

Indeed, one has

$$(x|x - w(x)) = \frac{1}{2}|x - w(x)|^2 \in \mathbf{Z}.$$

Consider the following subgroup of M_w :

$$M_w' = \{x \in M_w \mid \psi(x,y) = 1 \text{ for } any\ y \in M_w\} \tag{39}$$

As an exercise, we compute M_w' for the two previous examples.

---

* The authors of ref. 5,10 say that this modification is also required by modular invariance and the fact that one keeps the left-movers only.

Example 1 (cont).

$$M_w = \{x \in \underline{h} \mid (x|\gamma) \in \mathbf{Z}\}$$

$$(x|y - w(y)) = (x|\gamma)\,(y|\gamma)$$

Hence,

$$M_w{}' = M_w$$

Example 2 (cont).

$$M_w = \frac{1}{2}Q$$

$$(x|y - w(y)) = (x|2y) = \frac{1}{2}(\gamma|\gamma') \in \mathbf{Z}\ ,\ \gamma,\gamma' \in Q$$

This implies

$$(\gamma|\gamma') \in 2\mathbf{Z} \text{ for } any\ \gamma',$$

so that $\gamma \in 2Q$, and

$$M_w{}' = Q$$

Returning to the general case, from the definition (36) and (37,39) we see that

$$\psi(x \ mod\ M_w{}' ,y\ mod\ M_w{}') = \psi(x,y) \tag{40}$$

so that ψ is well-defined on the quotient

$$N_w = M_w/M_w{}' \tag{41}$$

One can prove that N_w is a finite (abelian) group whose order is a perfect square c_w^2. The integer c_w is called the defect of w. In example 1 above, $c_w=1$, while in example 2, $c_w=2^{d/2}$.

The final recipe for the space V in (33) is

$$V = \mathrm{C}(Q_0) \otimes \bar{V} \tag{42}$$

where $\mathrm{C}(Q_0)$ is the linear span of states $|k_0>$, and $\bar{V}$ is the space of an irreducible projective representation ρ of N_w which satisfies

$$\rho(x)\ \rho(y) = \psi(x,y)\ \rho(y)\ \rho(x) \tag{43}$$

and $\dim\bar{V} = c_w$. One can also define a two-cocycle ϵ_w by the conditions (10a-c), with the symmetry factor on the RHS of (10d) replaced by the map ψ. Then one gets

$$\rho(x)\ \rho(y) = \epsilon_w(x,y)\ \rho(x+y). \tag{44}$$

It is very instructive to look at this representation ρ in example 2.

Example 2 (end).

We have seen that $N_w = Q/2Q$. We look for a ρ s.t.

$$\rho(\alpha)\ \rho(\beta) = \tilde{\psi}(\alpha,\beta)\ \rho(\beta)\ \rho(\alpha)$$

where $\alpha,\beta \in Q/2Q$. The symmetry factor reduces to

$$\tilde{\psi}(\alpha,\beta) = (-1)^{(\alpha|\beta)}$$

so that $\epsilon_w = \epsilon$, the Frenkel-Kac cocycle. Now we construct $\rho(\alpha)$ in terms of γ matrices in some Clifford algebra. We restrict ourselves to the interesting case $g = E_8$.

As generators of Q we take first the simple roots $\{\alpha_1,\ldots,\alpha_8\}$. Their square length being 2, the Cartan matrix or the Dynkin diagram (see fig. 2) gives the scalar products $(\alpha_i | \alpha_j)$, the simple roots being indexed according to the numbering of vertices in the diagram.

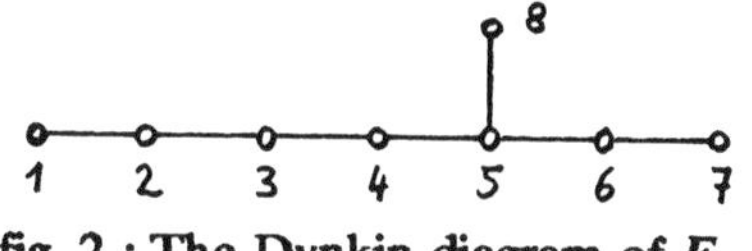

fig. 2 : The Dynkin diagram of E_8.

Adding one more generator α_0 satisfying

$$|\alpha_0|^2 = -1$$

$$(\alpha_0 \mid \alpha_j) = 0, \; j=1,\ldots,8$$

we get a lattice $Q \oplus \mathbf{Z}$ which is integer, odd, self-dual and Lorentzian. By a uniqueness theorem[11)] for this kind of lattices,

$$Q \oplus \mathbf{Z} = \mathbf{Z}^{1,8}$$

i.e. we have constructed a cubic Lorentzian lattice. This means there exist orthonormal generators $e_0, \ldots, e_8$ of $Q \oplus \mathbf{Z}$:

$$(e_i \mid e_j) = \eta_{ij} = diag(-++++++++)$$

Take the Clifford algebra with generators γ_i, $i=0,\ldots,8$ and relations :

$$\{\gamma_i \, , \gamma_j\} = 2\eta_{ij}$$

which can be written in this fancy way :

$$\gamma_i \, \gamma_j = (-1)^{[(e_i | e_j)^2 - |e_i|^2 |e_j|^2]} \, \gamma_j \, \gamma_i$$

For any lattice point

$$u = \sum_{i=0}^{8} n_i e_i \in \mathbf{Z}^{1,8}$$

we put

$$\gamma_u = \prod_{i=0}^{8} \gamma_i^{n_i}$$

Then one still has that

$$\gamma_u \, \gamma_v = (-1)^{[(u|v)^2 - |u|^2|v|^2]} \, \gamma_v \, \gamma_u \tag{45}$$

Since $Q \subset \mathbf{Z}^{1,8}$, the map ρ is just

$$\rho : \alpha \mapsto \gamma_\alpha \quad \text{for } \alpha \in Q$$

as the sign in the commutation relations (45) reduces to $\tilde{\psi}$ on Q. This argument can be found in ref. 12. The only thing which is missing in order to have an explicit construction of ρ is the relation between the two sets of generators $\{e_i\}$ and $\{\alpha_i\}$. You can check that the following linear combinations give the simple roots in terms of the orthonormal basis :

$$\alpha_1 = -3e_0 + 2e_1 + e_2 + \cdots + e_7 - e_8$$

$$\alpha_i = - e_{i-1} + e_i \qquad i=2,...,7.$$

$$\alpha_8 = e_0 - e_5 - e_6 - e_7$$

Two remarks : first one for finite group theory connaisseurs. ρ is a representation of the extraspecial 2-group 2_+^{1+8}. Second, the construction of ρ can easily be extended to one of the adjoint representation of E_8. See Frenkel, Lepowsky, and Meurman[3].

Back to the general case, we fix some notations for the oscillators. For each $j \in \{0,...,m-1\}$ let

$$\{ h_j^I \}, \ I=1,...,d_j \tag{46}$$

be an orthonormal basis of $\underline{h}_j$. The oscillators are :

$$h^I_{r+j/m}\ ,\ r \in \mathbf{Z}. \tag{47}$$

They satisfy the commutation relations :

$$\left[\ h^I_{r+j/m}\ ,\ h^J_{s+k/m}\ \right] = (r+j/m)\ \delta^{IJ}\ \delta_{r+s+(j+k)/m,0} \tag{48}$$

A more compact notation can be set up as follows. For an arbitrary vector $\alpha \in \underline{h}$, recall that α_j is its projection on the subspace $\underline{h}_j$. Let α^I_j be the coordinates relative to the basis (46). For $n \in \frac{1}{m}\mathbf{Z}$, there is a unique j between 0 and m-1 determined by :

$$j/m = n \ mod\ \mathbf{Z} \tag{49}$$

We define

$$\alpha(n) = \sum_{I=1}^{d_j} \alpha^I_j h^I_n \tag{50}$$

Then the commutation relations look simpler :

$$[\alpha(n),\beta(k)] = n(\alpha_j|\ \beta_l)\ \delta_{n,-k} \tag{51}$$

With these definitions, one can then write Virasoro algebra generators :

$$L^w_k = \frac{1}{2} \sum_{n \in \frac{1}{m}\mathbf{Z}} \sum_{i=1}^{d} :u_i(-n)u_i(n+k): \tag{52}$$

with $\{u_i\}$, $i=1,\dots,d$ an orthonormal basis of $\underline{h}$. These have the commutation relations:

$$[L^w_k, L^w_p] = (k-p)L^w_{k+p} + (\frac{d}{12}k^3+\mu k)\delta_{k,-p} \tag{53}$$

where

$$\mu = \frac{1}{2m^2}\sum_{j=1}^{m-1} j(m-j)d_j - \frac{d}{12}. \tag{54}$$

As in (17), we can decompose H_w into eigenspaces w.r.t.

$$L_0^w = \frac{1}{2}|p_0|^2 + \sum_{k>0} u_i(-k)u_i(k) \tag{55}$$

that is, we have

$$H_w = \bigoplus_{n \in 1/2m\ \mathbb{Z}_+} H_n . \tag{56}$$

The partition function of the twisted Hilbert space is then easily obtained :

$$dim_q H_w = c_w \frac{\sum_{k \in Q_0} q^{\frac{1}{2}|k|^2}}{\prod_{j=1}^{\infty}(1-q^{j/m})^{d_{(j \bmod m)}}} \tag{57}$$

The graphical picture of the Hilbert space is the same as in fig. 1, except for the fact that you have to think of the horizontal axes as Q_0 instead of Q. However in case $\underline{h}_0 = 0$, Q_0 is trivial and one has a Hilbert space consisting of a finite sum of Fock spaces (fig. 3). This possibility is realized for instance in example 2 above.

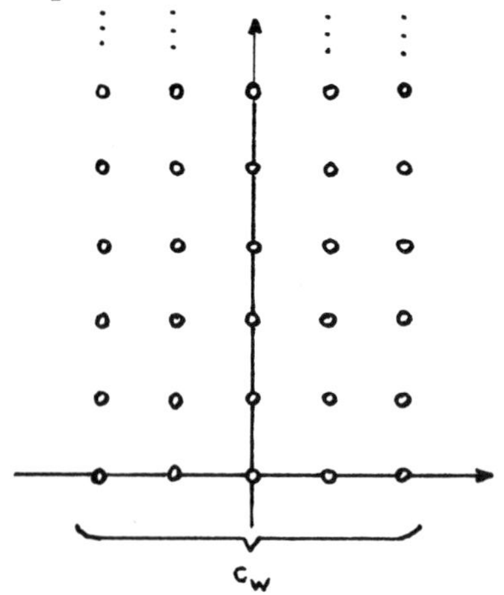

fig. 3 : The Hilbert space when $\underline{h}_0 = 0$.

4. THE TWISTED VERTEX OPERATOR.

We start by reviewing some basic facts about Q, the root lattice of a semi-simple simply-laced Lie algebra g. The set of roots is the subset

$$\Delta = \{ \alpha \in Q \mid |\alpha|^2 = 2 \}. \tag{58}$$

The Weyl group W is generated by the reflections r_α, $\alpha \in \Delta$:

$$r_\alpha(x) = x - (x|\alpha)\alpha \tag{59}$$

W is an invariant subgroup of Aut(Q), and

$$Aut(Q)/W = Aut(D) \tag{60}$$

where D is the Dynkin diagram of g.

The Twisted Vertex Operator is defined for $\alpha \in \Delta$ to be

$$V(\alpha,z) = :exp\left[- \sum_{n \in 1/m\mathbf{Z}^*} \frac{1}{n} z^{-n} \alpha(n) \right] : \otimes T_\alpha(z) \tag{61}$$

where $T_\alpha(z)$ is an operator acting in V - see (42) - whose definition requires a little bit more explanations. Consider the group

$$L_w = \{(\alpha,\beta) \in \underline{h}_0 \times \underline{h} \mid \beta - w(\beta) \in \alpha + Q\}. \tag{62}$$

Observe that

$$M_w = \{(0,\beta) \mid \beta - w(\beta) \in Q\} \subset L_w \tag{63}$$

$T_\alpha(z)$ is an operator belonging to a projective representation of L_w on V. This projective representation ρ' is irreducible and satisfies

$$\rho'(\gamma_1)\, \rho'(\gamma_2) = \Psi(\gamma_1,\gamma_2)\, \rho'(\gamma_2)\, \rho'(\gamma_1) \tag{64}$$

where the symmetry factor Ψ is the map

$$\Psi : L_w \times L_w \rightarrow \mathbf{C}^* \tag{65}$$

$$\Psi((\alpha_1,\beta_1),(\alpha_2,\beta_2)) = \exp 2\pi i\Theta \cdot \psi(\beta_1,\beta_2)$$

$$\Theta = \frac{1}{2}(\alpha_1|\,\alpha_2)+(\alpha_1|\,\beta_2)-(\alpha_2|\,\beta_1)$$

Ψ is bimultiplicative and alternating. By definition,

$$\rho' : (-\alpha_0,\beta) \longmapsto T_\alpha(z) \tag{66}$$

where α_0 is the projection of α on h_0 and

$$\beta = [\frac{m}{2}+\frac{1}{2\pi i}\log(z)]\alpha_0 - \frac{1}{m}\sum_{j=1}^{m-1} jw^j(\alpha). \tag{67}$$

Note that whenever $h_0 = 0$, many simplifications occur :

$$L_w = M_w,\ V = \bar{V},\ \rho' = \rho,\ \Psi = \psi.$$

Now we can formulate the fundamental result of ref. 6 (see also ref. 7). For the moment we restrict ourselves to $w \in W$ (this covers all the cases when $g = E_8$ as eq. (60) tells us).

Theorem

Expand in power series :

$$V(\alpha,z) = \sum_{n\in\frac{1}{m}\mathbf{Z}} V_n(\alpha)z^n .$$

Then for $\alpha \in \Delta$ the operators $V_n(\alpha)$, $\alpha(n)$ and the identity span a Lie algebra on H_w. This Lie algebra is isomorphic to $\hat{g}$, the affine Lie algebra corresponding to g. The representation on H_w is equivalent to the basic irreducible representation of level one of $\hat{g}$, i.e. the one we got using the Frenkel-Kac construction as explained in the introduction.

Note that if we would have taken $w \in Aut(Q)-W$, this would give us a construction of a twisted affine Lie algebra, in the sense of ref. 13. However in this paper we only study the untwisted case $w \in W$ which already has a very rich structure.

The zero modes $V_0(\alpha)$, $\alpha(0)=p_0$ span a finite-dimensional subalgebra g_0, which is a subalgebra of g and commutes with L_0^w :

$$[L_0^w, g_0] = 0. \tag{68}$$

This immediately implies that in the eigenspace decomposition (56), the H_n are representation spaces of g_0.

One can define an automorphism σ *of* g s.t. g_0 is the subalgebra of fixed points of σ and

$$\sigma(x) = w(x) \quad \text{for } x \in \underline{h} \tag{69}$$

$\underline{h}$ being identified with a Cartan subalgebra of g. σ is of order m or 2m if w is of order m.

The Frenkel-Kac construction corresponds to the trivial element w=1. It is also called the homogeneous construction. It is possible to describe explicitly the isomorphism between the homogeneous and the twisted construction. In fig. 4 we have depicted this isomorphism in a very simple but generic case. The horizontal axis is Q, the vertical axis the eigenvalues of L_0 (16). The interpretation of the dots is the same as in fig. 1. The intersection of the oblique hyperplanes with the paraboloid are the eigenspaces H_n (56) of the twisted construction.

Because of this non-trivial isomorphism, the twisted constructions can be used instead of the homogeneous one in the heterotic string, resulting in an equivalent theory. However the interpretation of the Hilbert space states will be different if one chooses L_0^w and g_0 to describe the mass

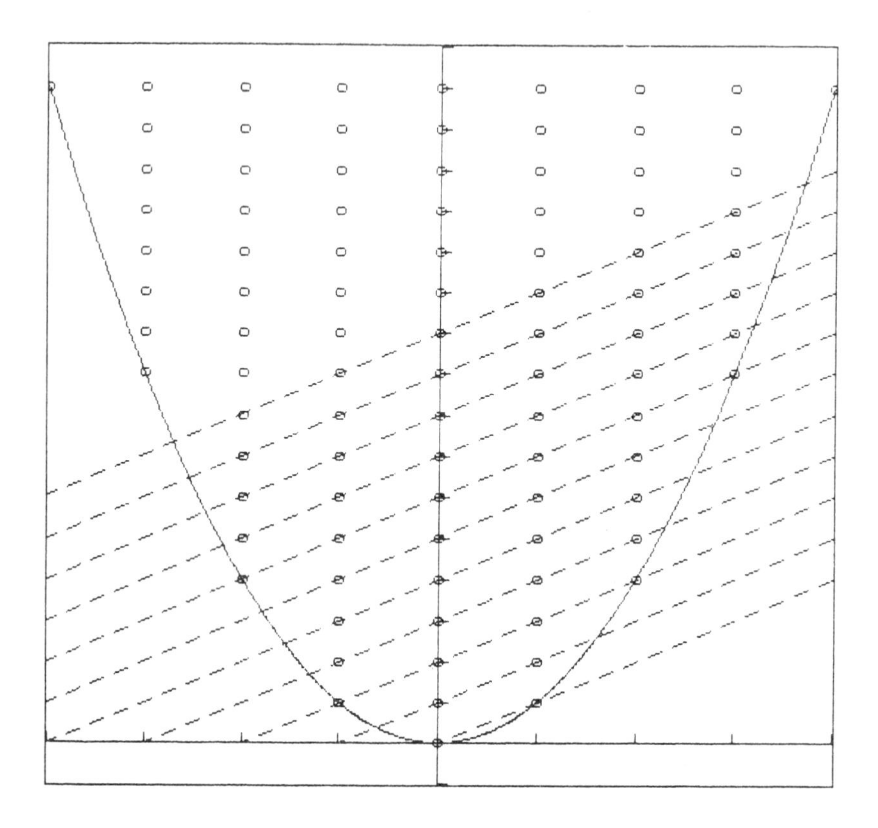

fig. 4 : The isomorphism between the twisted and the untwisted Hilbert spaces.

spectrum and the gauge quantum numbers instead of L_0 and g. This is a kind of non-trivial boson-boson equivalence. We can see it as a symmetry breaking from g to g_0.

A natural and important question is : given a choice of w, what is the subalgebra g_0 in the corresponding twisted construction ? This is the subject of the next paragraph.

5. DETERMINATION OF g_0.

It can be shown that

$$w \quad \text{and} \quad w'w(w')^{-1}$$

give the same constructions. Thus g_0 depends only on the conjugacy class of w in Aut(Q).

From now on we assume $g = E_8$, though our method is general. There are 112 conjugacy classes in the Weyl group of E_8. Consider the homogeneous construction of $\hat{g}$ and the corresponding version (3) of the Hilbert space, which is the basic level 1 representation. The weights of H have the general form[13) :

$$\lambda = \Lambda - \sum_{i=0}^{8} k_i \alpha_i, \quad k_i \in \mathbf{Z}_+ \tag{70}$$

where Λ is the highest weight.

The character is a multivariable generating function :

$$ch(H) = \sum_{\lambda} mult(\lambda)\, x_0^{k_0} x_1^{k_1} \cdots x_8^{k_8} \tag{71}$$

where the sum runs over all weights of H, and $mult(\lambda)$ is the multiplicity (degeneracy) of the weight space. A specialisation of the character is the single variable generating function $dim_q H(s)$ obtained by relating all variables in the character in a special way :

$$dim_q H(s) = ch(H)\Big|_{x_i = q^{s_i},\ i=0,\ldots,8} \tag{72}$$

where s is a sequence of non-negative integers :

$$s = (s_0, s_1, \ldots, s_8). \tag{73}$$

One says that (72) is the specialisation of type s. The vector whose components are the s_i gives the direction perpendicular to the hyperplanes in fig. 4 which can be seen as a graphical picture of the specialisation. By specialising the character, one adds multiplicities in the Hilbert space in a non-homogeneous (non-horizontal) way. But as we explained before, the family of oblique hyperplanes reflects precisely the isomorphism between the homogeneous and a twisted construction. Therefore by carefully selecting the type s of a specialisation, one should obtain the partition function (57) of the twisted construction. In other words (57) and (72) have to agree for some s :

$$dim_q H(s) = dim_q H_w. \tag{74}$$

Now for any type s one can associate an automorphism $\sigma(s)$ of g, and its invariant subalgebra $g_0(s)$. Let Σ be the set of conjugacy classes of automorphisms of g of order m or 2m. From the classification of automorphisms of finite order, it follows that this set is finite. We let $\sigma(s)$ run through all elements of Σ, and we will necessarily find the types s for which (74) holds. In the case of E_8 there is only one solution. In general there will be several ones related by an automorphism of the Dynkin diagram of the affine algebra $\hat{g}$, reflecting the fact that there are several distinct representations of level one.

Finally, we find that $g_0 = g_0(s)$ for a type s satisfying (74). Hence the problem of finding the invariant subalgebras g_0 for all conjugacy classes [w] has been reduced to an algorithmic procedure, which can be translated into computer programs. Some important examples of subalgebras g_0 obtained by this method for E_8 will be found in the next paragraph. In a separate paper[14)] we will develop in their full extent the arguments sketched above which lead us to the solution, together with the full list of subalgebras for the 112 conjugacy classes of the Weyl group of E_8.

6. RESULTS.

The conjugacy classes of the Weyl groups of all simple Lie algebras have been classified in a nice way by Carter[15)]. Each class has an associated Dynkin-like diagram, for example the diagram in fig. 5. Each vertex stands for a root, and the number of links between two of them is the absolute value of the scalar product of the corresponding roots. One then numbers the vertices in an arbitrary way, e.g. from left to right. After that one divides the set of vertices into two disjoint subsets A and B such that each one of them contains roots which are all mutually orthogonal, e.g. in the case of fig. 5 , $A = \{1,3,5\}$ $B = \{2,4\}$. Now let r_i be the reflection in the root corresponding to the vertex i. An element of a conjugacy class associated to the diagram is

$$w = \prod_{i \in A} r_i \cdot \prod_{i \in B} r_i .$$

In all cases but for a few exceptions the correspondance between the diagrams and the conjugacy classes is bijective. For the exceptions and the admissible diagrams for a given Weyl group see ref. 15.

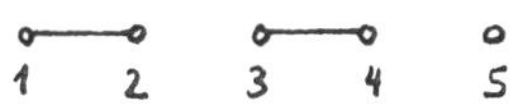

1 2 3 4 5

fig. 5 : The diagram of a conjugacy class in W.

We now give a few examples of conjugacy classes in the Weyl group of E_8 which correspond to "interesting" invariant subalgebras g_0, namely cases which incorporate most of the GUTs gauge groups and which are compatible with "phenomenology".

1) [w]= o

This corresponds to example 1 above. One finds

$$G_0 = E_7 \times U(1)$$

G_0 is a Lie group with Lie algebra g_0 in physicists notation.

2) [w]= o o o o o o o o

This case corresponds to example 2, and

$$G_0 = O(16)$$

but of course this was known previously[3].

3) [w]= o o o o o o

The order of w is m=2,

$$d_0 = 2, \quad d_1 = 6, \quad c_w = 4,$$

$$G_0 = O(10) \times O(6)$$

4) [w]= o o o o

The order is m=2,

$$d_0 = d_1 = 4.$$

This diagram is one of the few which corresponds to *two* conjugacy classes. For one of them

$$G_0 = E_7 \times SU(2), \quad c_w = 2$$

and for the other

$$G_0 = SU(8) \times U(1), \quad c_w = 1.$$

The first one can be found in ref. 5.

5) [w]= o—o o—o o—o o—o

The order is m=3,

$$d_0 = 0, \quad d_1 = d_2 = 4, \quad c_w = 9,$$

$$G_0 = SU(9).$$

6) [w]= o—o o—o o—o

The order is m=3,

$$d_0 = 2, \quad d_1 = d_2 = 3, \quad c_w = 3,$$

$$G_0 = E_6 \times SU(3).$$

7) [w]= o—o—o—o o—o—o—o

The order is m=5,

$$d_0 = 0, \quad d_j = 2, \quad j=1,\ldots,4, \quad c_w = 5,$$

$$G_0 = SU(5) \times SU(5).$$

8) [w]=

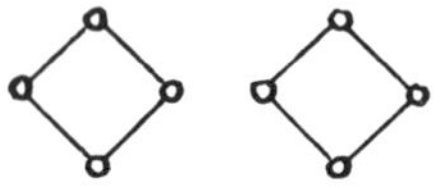

The order is m=4,

$$d_0 = d_2 = 0, \quad d_1 = d_3 = 4, \quad c_w = 4,$$

$$G_0 = O(10) \times O(6).$$

7. FURTHER PROBLEMS.

The Vertex Operator $V(\alpha,z)$ in (61) is the essential ingredient of the physical Vertex Operator for the emission of an untwisted string state, e.g. gauge boson, graviton ..., from the twisted string.

To obtain a consistent new interacting string theory, one must consider simultaneoulsly both twisted and untwisted strings. Therefore we need Vertex Operators for the emission of twisted string states from an (un)twisted string. The authors of ref. 5,16 are working in this direction.

To compute amplitudes involving the Vertex (61), it would be nice to have a description of the representation ρ *of* N_w in the general case, as explicit as it is in example 2. This is an interesting group-theoretical problem.

Our last comment is that even without trying to build new string theories, one can replace the homogeneous (untwisted) Hilbert space H in the heterotic string by a twisted one, H_w for $w \in W$. Since they are isomorphic, we get hundreds of equivalent versions of the heterotic string. This fact can be useful, e.g in relation with fermionization and for the study of modular invariance and supersymmetry. But what is the physical meaning of this large number of boson-boson equivalence which we wanted to emphasize here ? Today we can do no more than suggest that there should be a symmetry responsible for that which is hidden by the present formulation of the theory.

Acknowledgements. D.A. would like to thank V. Rittenberg and the members of the organising commitee of the Workshop for inviting him to deliver this talk. He also thanks J.P. Eckmann, S. Kamphorst and A. Malaspinas for their patience while explaining how to use the computer efficiently.
I.R. appreciated the warm hospitality of the Department of Theoretical Physics at the University of Geneva.
This work was partially supported by the Swiss National Science Foundation, and by the Conselho Nacional de Desenvolvimento Cientifico e Tecnologico (CNPq, Brazil).

References.

1. I. Frenkel and V. Kac, Inv. Math. 62 , 23 (1980).
   G. Segal, Comm. Math. Phys. 80 , 301 (1981).

2. D. Gross, J. Harvey, E. Martinec and R. Rohm, Phys. Rev. Lett. 54 , 502 (1985); Nucl. Phys. B256 , 253 (1985); Nucl. Phys. B267 , 75 (1986).

3. I. Frenkel, J. Lepowsky and A. Meurman, in Vertex Operators in Mathematics and Physics, proceedings of a conference in Berkeley, nov. 1983, MSRI publications #3, Springer-Verlag 1984; in Contemporary Mathematics vol. 45, 1985, published by the AMS; Proc. Natl. Acad. Sci. USA 81 , 3256 (1984).

4. E. Corrigan, Phys. Lett. 169B , 259 (1986) and Durham University preprint DTP-85/21.

5. L. Dixon, J Harvey, C. Vafa and E. Witten, Nucl. Phys. B261 , 651 (1985); Nucl. Phys. B274 , 285 (1986).

6. V. Kac and D. Peterson in the proceedings of a conference on Anomalies, Geometry and Topology in Argonne, Chicago, ed. A. White, World Scientific, 1985.

7. J. Lepowsky, Proc. Natl. Acad. Sci. USA 82 , 8295 (1985).

8. P. Goddard, D. Olive and A. Schwimmer, Vertex Operators for Non-simply-laced Algebras. Cambridge preprint, august 1986.
   D. Bernard and J. Thierry-Mieg, Meudon preprint, august 1986.

9. S. Roy and V. Singh, Tata Institute preprints TIFR/TH/85-20 and 21.

10. J. Harvey in Unified String theories, proceedings of a workshop in Santa Barbara, august 1985, M. Green and D. Gross eds., World Scientific, 1986.

11. J.P. Serre, Cours d'arithmétique, Presses Universitaires de France, 1970.

12. P. Goddard and D. Olive in Vertex Operators in Math. and Phys., op. cit.

13. V. Kac, Infinite-dimensional Lie algebras, 2nd ed., Cambridge University press, 1985.

14. D. Altschüler, Ph. Béran, J. Lacki and I. Roditi, in preparation.

15. R. Carter, Compositio Mathematica 25 , 1 (1972).

16. L. Dixon, Ph. D. thesis, Princeton University, July 1986.
L. Dixon, D. Friedan, E. Martinec and S. Shenker, Princeton-Chicago preprint, 1986.
S. Hamidi and C. Vafa, Caltech preprint CALT-86-1349.

UNITARITY and MODULAR INVARIANCE

L. Dolan

The Rockefeller University, New York, N.Y. 10021
USA

ABSTRACT

A connection between modular invariance and perturbative unitarity in dual string models is discussed. In particular, a compactified ten-dimensional string which is invariant under a modular subgroup of finite index, is rewritten as a modular invariant theory, a step which picks out the spectrum on which this model is unitary. This analysis led to the mechanism for the non-abelian compactification of type II superstrings.

1. INTRODUCTION

In the light-cone gauge of dual string theory, the Fock space contains only physical states. The condition that the S-matrix of a theory be unitary gives an equation for the imaginary part of the N-particle scattering amplitudes. For example, unitarity at the one-loop level in string perturbation theory relates the discontinuity across a given cut of a one-loop amplitude to a particular product of tree amplitudes. One can therefore investigate perturbative unitarity by calculating the absorptive part of a one-loop amplitude directly, and then comparing it with the value required by the unitarity equation. As a review, this procedure is illustrated in Section 2 for the planar graph of the open Veneziano string. For oriented closed strings, the one-loop amplitudes are usually defined as integrals over a fundamental region of the modular group. This integration region, while motivated by the modular invariance of the integrand, is actually fixed by the requirements of unitarity and analyticity. For example, the $\mathrm{Im}\tau\to\infty$ limit of the fundamental region gives the complete contribution to the imaginary part corresponding to the lowest threshold. An integration over more than one fundamental region would therefore give (non-unitary) additional copies of the discontinuity. Closed string amplitudes in some cases appear to be invariant only under a subgroup of the modular group. When the fundamental region of the subgroup covers the fundamental region of the modular group a finite number of times, these amplitudes can be rewritten as modular invariant expressions, and their unitarity properties can then be understood in the standard way. This rewriting in general changes the integrand, thereby adding contributions to the absorptive parts. This signals an additional sector of states which must be included for the interacting string to satisfy perturbative unitarity. In Section 3, this procedure is used as a trick to discover how type II superstrings can be compactified to incorporate a non-abelian gauge group in four dimensions. This review is meant to give some insight into how modular or sub-modular invariance leads to unitarity, at least in one-loop. For simplicity, the equations of Section 3 are thus given for the gauge group $SU(2)^6$. This group can be generalized to any dimension 18 semi-simple group, i.e. SU(3)xSO(5) or SU(2)xSU(4), these being large enough to contain the standard group SU(3)xSU(2)xU(1) and thus providing a mechanism by which the type II string may be phenomenologically viable. Section 3 is based on references [1] and [2].

2. PERTURBATIVE UNITARITY IN STRINGS

2.1 Tree and One-loop Amplitudes and Their Singularities

The only singularities of a string tree diagram are poles, since it is a zero-width approximation. The imaginary part of a tree amplitude is therefore a sum of delta functions, each multiplied by the residue at the pole (Cauchy's theorem). In order to fix notation, we define the vertex operator for the emission of tachyons in the open Veneziano string, the propagator, and the four-point tree amplitude for external tachyons:

$$V(k,z) = z^{\alpha' k^2} e^{\sqrt{2\alpha'}\, k\cdot\sum_{n=1}^{\infty} \frac{A_{-n}}{n} z^n} e^{ik\cdot x} z^{2\alpha' k\cdot p} e^{-\sqrt{2\alpha'}\, k\cdot\sum_{n=1}^{\infty} \frac{A_n}{n} z^{n}} \tag{1}$$

where $\alpha' k^2=1$,

$$\Delta = \frac{1}{\alpha' p^2+\alpha' m^2} = \int_0^1 dz\; z^{\alpha' p^2 + N - 2} \tag{2}$$

and

$$\begin{aligned} A(1234) &= \langle 0;-k_1|\, V(k_2)\, \Delta\, V(k_3)\, |0;k_4\rangle \\ &= \delta^{26}(k_1+k_2+k_3+k_4)\, \alpha' g^2\, \frac{\Gamma(-\alpha' s-1)\,\Gamma(-\alpha' t-1)}{\Gamma(\alpha' s-\alpha' t-2)} \\ &= \delta^{26}(\Sigma k)\;\; A(s,t)\;. \end{aligned} \tag{3}$$

For $\alpha' s<-1$ and $\alpha' t<-1$, an integral representation for (3) is

$$A(s,t) = \int_0^1 dz\, z^{-2-\alpha' s}(1-z)^{-2-\alpha' t}. \tag{4}$$

From Cauchy's theorem, we can write a dispersion relation for A(s,t) at fixed t:

$$A(s,t) = \lim_{\epsilon\to 0} \frac{1}{\Pi}\int ds' \frac{\mathrm{Im}A(s',t)}{s'-s-i\epsilon}$$

$$\mathrm{Im}A(s,t) = \frac{\Pi}{\alpha'}\sum_{n=0}^{\infty} \delta\left(s-\frac{n-1}{\alpha'}\right)\frac{1}{n!}(2+\alpha' t)\ldots(n+1+\alpha' t) \tag{5}$$

Thus the absorptive part of A(1234) is an infinite sum of delta functions. This is the "discontinuity at the poles". Also, by expanding the integrand in (3) around z=0, we can express A(s,t) as an infinite sum of simple poles:

$$A(s,t) = \sum_{n=0}^{\infty} \frac{1}{\alpha'}\frac{1}{n!}(2+\alpha' t)\ldots(n+1+\alpha' t)\,\frac{1}{s-\left(\frac{n-1}{\alpha'}\right)}\ . \tag{6}$$

Here $s=-(k_1+k_2)^2$, $t=-(k_1+k_3)^2$.

The one-loop amplitudes have poles and cuts. The four-point one-loop planar graph for external tachyons is

$$A_{\text{planar loop}}(1234) = \int dp\ \mathrm{tr}\Delta V(1)\Delta V(2)\Delta V(3)\Delta V(4)$$

$$= (\alpha' g)^4 \int dp \int_0^1 \prod_{I=1}^{4} dz_I z_I^{\alpha' p_I^2}\, \omega^{-2}(f(\omega))^{-24}$$

$$\prod_{I<J}\prod_{n=1} \frac{\left(1-\omega^{n-1}\frac{\rho_J}{\rho_I}\right)\left(1-\omega^{n}\frac{\rho_I}{\rho_J}\right)}{(1-\omega^n)^2}^{2\alpha' k_I\cdot k_J}\ . \tag{7}$$

Here $\rho_I = z_1 \ldots z_I$, $\omega=\rho_4= e^{-2\Pi\tau}$.

In general, the absorptive part of the one-loop amplitude is an infinite number of step functions, each multiplied by a product of tree amplitudes, a result of the two-particle intermediate states. To calculate directly the imaginary part which would correspond to the two-tachyon threshold in the s channel, it is sufficient to consider the $\omega\to 0$ limit of the integrand in (7) since the discontinuity of the loop integral as a function of s at fixed t is given by that region of integration where the derivative of the integral with respect to s is infinite, i.e.

$$\text{disc } A_{\text{planar}}(1234)_{\text{loop}} = \text{disc } (\alpha' g)^4 \int dp \int \prod_{I=1}^{4} dz_I \lim_{\substack{z_1, z_3 \to 0 \\ z_2, z_4 \text{ fixed}}} (z_I{}^{\alpha' p_I^2} \omega^{-2}$$

$$(f(\omega))^{-24} \prod_{I<J} \prod_{n=1}^{\infty} \frac{(1-\omega^{n-1} \frac{\rho_J}{\rho_I})(1-\omega^n \frac{\rho_I}{\rho_J})^{2\alpha' k_I \cdot k_J}}{(1-\omega^n)^2})$$

$$= \text{disc } (\alpha' g)^4 \int dp \int dz_1 dz_3 \; z_1{}^{\alpha' p^2 - 2} \; z_3{}^{\alpha'(p-k_1-k_2)^2 - 2}$$

$$\int dz_2 dz_4 \; z_2{}^{\alpha'(p-k_1)^2 - 2} \; z_4{}^{\alpha'(p-k_1-k_2-k_3)^2} ((1-z_2)(1-z_4))^{2\alpha' k_1 \cdot k_2}$$

$$= \text{disc } (\alpha' g)^4 \int dp \frac{1}{\alpha' p^2 - 1} \; \frac{1}{\alpha'(p-k_1-k_2)^2 - 1}$$

$$\int dz_2 \, dz_4 \; z_2{}^{-2\alpha' p \cdot k_1 + \alpha' p^2 - 1} z_4{}^{-2\alpha' p \cdot (k_3+k_2+k_1) + \alpha' p^2 - 1}$$

$$[(1-z_2)(1-z_4)]^{2\alpha' k_1 \cdot k_2} \tag{8a}$$

$$= (\alpha' g)^4 (-(2\Pi)^2) \int d^{26} p \; \Theta(p^0) \; \Theta(p^0 - (k_1+k_2)^0) \; \delta(\alpha' p^2 - 1) \; \delta(\alpha'(p-(k_1+k_2))^2 - 1)$$

$$\int dz_2 dz_4 [(1-z_2)(1-z_4)]^{2\alpha' k_1 \cdot k_2} \; z_2{}^{-2\alpha' p \cdot k_1 + \alpha' p^2 - 1}$$

$$z_4{}^{-2\alpha' p \cdot (k_3+k_2+k_1) + \alpha' p^2 - 1} \tag{8b}$$

$$= (\alpha')^2 (-(2\Pi)^2) \int d^{26} p \; d^{26} \ell \; \delta^{26}(k_1+k_2-p-\ell) \; \Theta(p^0) \; \Theta(\ell^0) \delta(\alpha' p^2 - 1) \delta(\alpha' \ell^2 - 1)$$

$$\int dz_2 dz_4 [(1-z_2)(1-z_4)]^{2\alpha' k_1 \cdot k_2} \; z_2{}^{-2\alpha' p \cdot k_1} \; z_4{}^{2\alpha' p \cdot k_4} \tag{8c}$$

$$= (\alpha')^2 (-(2\Pi)^2) \int d^{26} p \; d^{26} \ell \; \delta^{26}(k_1+k_2-p-\ell) \; \Theta(p^0) \; \Theta(\ell^0) \delta(\alpha' p^2 - 1) \delta(\alpha' \ell^2 - 1)$$

$$A_{\text{tree}}(-E_p, -\vec{p}; 1, 2; -E_\ell, -\vec{\ell}) \;\; A_{\text{tree}}(\ell; 3, 4 : p). \tag{8d}$$

Equation (8b) follows from (8a), with use of the string analog of the Cutkosky prescription, which evaluates the imaginary part of the integral $A_{\text{loop}}(1243)$ as a function of the external momentum by replacing the two propagators inside the integrand in (8a) by

$$\frac{1}{\alpha' p^2 + \alpha' m^2} \to 2\Pi i \; \Theta(p^0) \; \delta(\alpha' p^2 + \alpha' m^2) \tag{9}$$

Equation (8d) follows from (8c) together with (3) and (4).

2.2 The Unitarity Equation

The unitarity equation is derived from the condition that the S-matrix satisfies $S^{\dagger}S=1$. Define the transition matrix R by $S=1+iR$. Then $S^{\dagger}S=1$ implies $R-R^{\dagger}=iR^{\dagger}R$, i.e.

$$\langle out|R|in\rangle - \langle out|R^{\dagger}|in\rangle = i\,\langle out|R^{\dagger}R|in\rangle \tag{10}$$

When the external particles are spinless, $\langle out|R^{\dagger}|in\rangle = \langle out|R|in\rangle^{*}$. Then Equation (10) becomes

$$2\mathrm{Im}\langle out|R|in\rangle = \langle out|R^{\dagger}R|in\rangle = \sum_j \langle out|R^{\dagger}|j\rangle\,\langle j|R|in\rangle \tag{11}$$

To make contact with the string amplitudes, we define a T-matrix by

$$\langle out|R|in\rangle = (2\Pi)^D\,\delta^D(k_{out}-k_{in})\,\langle out|T|in\rangle. \tag{12}$$

Then from (11),

$$\begin{aligned} 2\mathrm{Im}\langle out|T|in\rangle &= (2\Pi)^D \sum_j\langle out|T^{\dagger}|j\rangle\,\langle j|T|in\rangle\,\delta^D(k_{out}-k_j) \\ &= (2\Pi)^D(\alpha')^2\int d^Dp\,d^D\ell\,\langle out|T^{\dagger}|\ell+p\rangle\,\langle \ell+p|T|in\rangle \\ &\qquad \Theta(p^0)\Theta(\ell^0)\,\delta(\alpha'p^2-1)\delta(\alpha'\ell^2-1)\delta^D(\ell+p-k_{in}) \\ &\qquad + \text{sum over higher energy intermediate states} \end{aligned} \tag{13}$$

Equation (13) is the unitarity equation. In order to check the unitarity of the planar loop given in (8), we identify on the left side of (13):

$$\langle out|T|in\rangle = \langle -k_3-k_4|T|k_1+k_2\rangle = A_{loop}(1234)$$

and on the right side of (13):

$$\langle out|T|\ell+p\rangle = A_{tree}(p,\ell,3,4) = A_{tree}(\ell,3,4,p)\ , \text{ etc} \tag{14}$$

Then Equation (13) is equivalent to (8d).

For open strings, there are further one-loop contributions from the non-planar, oriented and non-oriented graphs. The discontinuities of these diagrams are given by products of the twisted trees, ie. the complete tree amplitude corresponding to (3) is

$$T(1234) = A(1234) + A(1324) + A(1243). \tag{15}$$

In addition, the well-known presence of bound state poles in the non-planar oriented graphs forces the coupling of open strings to closed strings for a consistent interacting theory. The description of perturbative unitarity presented here can be straightforwardly extended to theories of closed oriented strings.

The first quantized path integral analysis of string amplitudes of the interacting string picture is a manifestly unitary formualtion and allows one to show that all amplitudes, with arbitrary numbers of external particles and arbitrary types of external states satisfy (13). We have given this rather pedestrian review in this section, in order to show how considerations of unitarity led to a consistent compactification of type II superstrings.

3. THE COMPACTIFIED TYPE II SUPERSTRING

We first consider the compactified superstring x Neveu-Schwarz model [1,2]. The degrees of freedom are:

$$X^{\hat{i}}(\tau-\sigma) = \frac{x^{\hat{i}}}{2} + \alpha' p^{\hat{i}}(\tau-\sigma) + \frac{i\sqrt{2\alpha'}}{2} \sum_{n\neq 0} \frac{1}{n} A_n^{\hat{i}} e^{-2in(\tau-\sigma)} \tag{16}$$

$$X^{I}(\tau-\sigma) = \bar{x}^{I} + 2\alpha' \bar{p}^{I}(\tau-\sigma) + \frac{i\sqrt{2\alpha'}}{2} \sum_{n\neq 0} \frac{1}{n} A_n^{I} e^{-2in(\tau-\sigma)} \tag{17}$$

$$S^{a}(\tau-\sigma) = \sum_{n} S_n^{\ a} e^{-2in(\tau-\sigma)} \tag{18}$$

and

$${}^{B}X^{\hat{i}}(\tau+\sigma) = \frac{x^{\hat{i}}}{2} + \alpha' p^{\hat{i}}(\tau+\sigma) + \frac{i\sqrt{2\alpha'}}{2} \sum_{n\neq 0} \frac{1}{n} A_n^{\hat{i}} e^{-2in(\tau+\sigma)} \tag{19}$$

$$X^{I}(\tau+\sigma) = \tilde{x}^{I} + 2\alpha' \tilde{p}^{I}(\tau+\sigma) + \frac{i\sqrt{2\alpha'}}{2} \sum_{n\neq 0} \frac{1}{n} A_n^{I} e^{-2in(\tau+\sigma)} \tag{20}$$

$$b^{i}(\tau+\sigma) = \sum_{s} b_s^{i} e^{-2is(\tau+\sigma)} \ . \tag{21}$$

Here $i = \{\hat{i}=1,2;\ I=1,\ldots 6\}$.

The mass operator is

$$\alpha' m_4 = N + \tilde{N} - \frac{1}{2} + \alpha'(\tilde{p}^{\ I})^2 + \alpha'(\bar{p}^{I})^2 \tag{22}$$

The left-right gauge constraint is

$$\tilde{N} + \alpha'(\tilde{p}^{\ I})^2 = N + \alpha'(\bar{p}^{I})^2 - \frac{1}{2} \tag{23}$$

where

$$N = \sum_{n=1}^{\infty} A_{-n}^{i} A_{n}^{i} + \sum_{n=1}^{\infty} \frac{n}{2} S_{-n}\gamma^{-} S_{n} \tag{24a}$$

$$\tilde{N} = \sum_{n=1}^{\infty} \tilde{A}_{-n}^{i} \tilde{A}_{n}^{i} + \sum_{n=1}^{\infty} \frac{s}{2} b_{-s}^{i} b_{s}^{i} \tag{24b}$$

and
$$\sqrt{2\alpha'}\, \bar{p}^{I} = \Sigma N^{L} \delta_{L}^{I} \; , \; \sqrt{2\alpha'}\, \tilde{p}^{I} = \Sigma N^{L} \delta_{L}^{I} \; . \tag{25}$$

In four dimensions, the field content of the massless sector is

$$|\hat{i}>, | I>, |a> \quad x \quad b^{\hat{i}}_{-\frac{1}{2}}| 0> \tag{26a}$$

$$|\hat{i}>, |I>, |a> \quad x \quad b^{I}_{-\frac{1}{2}}| 0> \; , \; |p_{1}{}^{I}> \; . \tag{26b}$$

Equations (26a,b) are the N=4 supergravity multiplet and the N=4 supersymmetric Yang-Mills multiplet in SU(6) repectively.

The vertex operators and propagator and the tree amplitude for this model are given in [1,2]. The four-point one-loop amplitude for external charged vector mesons is given by

$$A_{\rm loop}(1234) = \epsilon_1^{\hat{i}} \epsilon_2^{\hat{j}} \epsilon_3^{\hat{k}} \epsilon_4^{\hat{\ell}} \int dp \; {\rm tr}\; \Delta_c W^{\hat{i}}(1) \Delta_c W^{\hat{j}}(2) \Delta_c W^{\hat{k}}(3) \Delta_c W^{\hat{\ell}}(4) \tag{27a}$$

$$= \left(\frac{\alpha' g}{2\Pi}\right)^{4} \frac{1}{16} \cdot K \frac{(2\Pi)^8}{(\alpha')^2} \int_{F_{SM}} d^2\tau \; ({\rm Im}\tau)^{-2}\; g(\tau) \tag{27b}$$

where K is the superstring kinematic factor, and the integrand $g(\tau)$ is given in (29) and is invariant under the theta subgroup Γ_Θ of modular transformations generated by $\tau \rightarrow \tau+2$ and $\tau \rightarrow -1/\tau$, and F_{SM} is a fundamental region of this subgroup (F_{SM}: $-1<{\rm Re}\tau<1$, ${\rm Im}\tau>0$, $|\tau|>1$.) Since this fundamental region F_{SM} is a finite union of images of F under Γ_Θ (where F is a fundamental region of the full modular group Γ, F: $-1/2<{\rm Re}\tau<1/2$, ${\rm Im}\tau>0$, $|\tau|>1$), we can rewrite (27) using

$$\int_{F_{SM}} d^2\tau \; ({\rm Im}\tau)^{-2}\; g(\tau) = \int_{F} d^2\tau \; ({\rm Im}\tau)^{-2} [g(\tau)+g(\tau+1)+g(-1/\tau+1)] . \tag{28}$$

From (27a), we find

$$\int d^2\tau (\mathrm{Im}\tau)^{-2} g(\tau) = \int d^2\tau \left(\frac{-2\pi}{\ell n\,|w|}\right)^2 \int \prod_I^3 d^2\nu_I \prod_{I<J} (\chi_{IJ})^{\alpha' \hat{k}_I^i \hat{k}_J^i} (f(\bar{w}))^{-8}$$

$$\bar{w}^{-1/2} \prod_{s=1/2}^{\infty} (1+\bar{w}^s)^8 \sum_{\sqrt{2\alpha'}\bar{p}^I \epsilon Z^6} w^{\alpha' (\bar{p}^I)^2} \prod_{I<J} (\Psi_{IJ})^{2\alpha' k_I^K k_J^K}$$

$$\sum_{\sqrt{2\alpha'}\tilde{p}^I \epsilon Z^6} (e^{\alpha' \ln\bar{w} [\tilde{p}^I - \sum_{m=1}^{4} \frac{\ln\bar{z}_m}{\ln\bar{w}} Q_m^I]^2})$$

$$[\tilde{k}_1 \cdot \tilde{k}_2\, \tilde{k}_3 \cdot \tilde{k}_4 \chi_{43}^+ \chi_{21}^+ - \tilde{k}_1 \cdot \tilde{k}_3\, \tilde{k}_2 \cdot \tilde{k}_4 \chi_{31}^+ \chi_{42}^+ + \tilde{k}_1 \cdot \tilde{k}_4\, \tilde{k}_3 \cdot \tilde{k}_2 \chi_{41}^+ \chi_{32}^+]. \tag{29a}$$

Using transformation properties of the Jacobi theta functions, we have from (29a),

$$\int d^2\tau (\mathrm{Im}\tau)^{-2} g(\tau+1) = \int d^2\tau \left(\frac{-2\Pi}{\ln|w|}\right)^2 \int \prod_I^3 d^2\nu_I \prod_{I<J} (\chi_{IJ})^{\alpha' \hat{k}_I^i \hat{k}_J^i} (f(\bar{w}))^{-8}$$

$$\cdot [-\bar{w}^{-1/2} \prod_{s=1/2}^{\infty} (1-\bar{w}^s)^8 \sum_{\sqrt{2\alpha'}\bar{p}^I \epsilon Z^6} w^{\alpha' (\bar{p}^I)^2} (-1)^{2\alpha' (\bar{p}^I)^2}] \prod_{I<J} (\Psi_{IJ})^{2\alpha' k_I^K k_J^K}$$

$$\sum_{\sqrt{2\alpha'} \tilde{p}^I \epsilon Z^6} (e^{\alpha' \ln\bar{w} [\tilde{p}^I - \sum_{m=1}^{4} \frac{\ln\bar{z}_m}{\ln\bar{w}} Q_m^I]^2} (-1)^{2\alpha' (\tilde{p}^I)^2})$$

$$\cdot [\tilde{k}_1 \cdot \tilde{k}_2\, \tilde{k}_3 \cdot \tilde{k}_4 \tilde{\chi}_{43}^+ \tilde{\chi}_{21}^+ - \tilde{k}_1 \cdot \tilde{k}_3\, \tilde{k}_2 \cdot \tilde{k}_4 \tilde{\chi}_{31}^+ \tilde{\chi}_{42}^+ + \tilde{k}_1 \cdot \tilde{k}_4\, \tilde{k}_2 \cdot \tilde{k}_3 \tilde{\chi}_{41}^+ \tilde{\chi}_{32}^+] \tag{29b}$$

and

$$\int d^2\tau (\mathrm{Im}\tau)^{-2} g(-1/\tau +1) = \int d^2\tau \left(\frac{-2\Pi}{\ln|w|}\right)^2 \int \prod_I^3 d^2\nu_I \prod_{I<J} (\chi_{IJ})^{\alpha' \hat{k}_I^i \hat{k}_J^i} (f(\bar{w}))^{-8}$$

$$\cdot [-\prod_{n=1}^{\infty} (1+\bar{w}^n)^8 16 \sum_{\sqrt{2\alpha'}\bar{p}^I \epsilon (Z+1/2)^6} w^{\alpha' (\bar{p}^I)^2} \prod_{I<J} (\Psi_{IJ})^{2\alpha' k_I^K k_J^K}]$$

$$\cdot \sum_{\sqrt{2\alpha'} \tilde{p}^I \epsilon (Z+1/2)^6} (e^{\alpha' \ln\bar{w} [\tilde{p}^I - \sum_{m=1}^{4} \frac{\ln\bar{z}_m}{\ln\bar{w}} Q_m^I]^2}$$

$$[\tilde{k}_1\cdot\tilde{k}_2\ \tilde{k}_3\cdot\tilde{k}_4\hat{\tilde{\chi}}^{+}_{43}\tilde{\tilde{\chi}}^{+}_{21} - \tilde{k}_1\cdot\tilde{k}_3\ \tilde{k}_2\cdot\tilde{k}_4\hat{\tilde{\chi}}^{+}_{31}\tilde{\tilde{\chi}}^{+}_{42}$$
$$+\tilde{k}_1\cdot\tilde{k}_4\ \tilde{k}_2\cdot\tilde{k}_3\hat{\tilde{\chi}}^{+}_{41}\tilde{\tilde{\chi}}^{+}_{32}]. \tag{29c}$$

We now identify the contribution of Equation(29a,b) as

$$\left(\frac{\alpha' g}{2\Pi}\right)^4 \frac{1}{16\alpha'^2}\ K\ (2\Pi)^8 \int_F d^2\tau\ (\mathrm{Im}\tau)^{-2}\ \left(\frac{g(\tau)+g(\tau+1)}{2}\right)$$

$$=\epsilon_1^{\hat{i}}\epsilon_2^{\hat{j}}\epsilon_3^{\hat{k}}\epsilon_4^{\hat{\ell}} \int dp\ \mathrm{tr}\ \left(\frac{1+(-1)^P}{2}\right)\Delta_c W^{\hat{i}}(1)\ \left(\frac{1+(-1)^P}{2}\right)\Delta_c W^{\hat{j}}(2)$$
$$\cdot\left(\frac{1+(-1)^P}{2}\right)\Delta_c W^{\hat{k}}(3)\ \left(\frac{1+(-1)^P}{2}\right)\Delta_c W^{\hat{\ell}}(4) \tag{30}$$

where $P = \sum_{s=1/2}^{\infty} b^i_{-s}b^i_s\ -1 + 2\alpha'(p^I)^2 + 2\alpha'(\tilde{p}^I)^2$.

Therefore $g(\tau)$ + $g(\tau+1)$ is a projection onto states with $(-1)^P=1$. This projection in fact does not eliminate any states from the original SSxNS spectrum since $(-1)^P=1$ is automatically satisfied by the left-right constraint (23).

Similarly, Eq.(29c) is identified as

$$\left(\frac{\alpha' g}{2\Pi}\right)^4 \frac{1}{16\alpha'^2}\ K\ (2\Pi)^8 \int_F d^2\tau\ (\mathrm{Im}\tau)^{-2}\ \frac{g(-1/\tau+1)}{2}$$

$$= -\epsilon_1^{\hat{i}}\epsilon_2^{\hat{j}}\epsilon_3^{\hat{k}}\epsilon_4^{\hat{\ell}}\ 1/2\int dp\ \mathrm{tr}\ \left(\frac{1+(-1)^{\overline{P}}}{2}\right)\Delta\overline{W}^{\hat{i}}(1)\ \left(\frac{1+(-1)^{\overline{P}}}{2}\right)\Delta\overline{W}^{\hat{j}}(2)$$
$$\left(\frac{1+(-1)^{\overline{P}}}{2}\right)\Delta\overline{W}^{\hat{k}}(3)\ \left(\frac{1+(-1)^{\overline{P}}}{2}\right)\Delta\overline{W}^{\hat{\ell}}(4) \tag{31}$$

where $(-1)^{\overline{P}} = 2^4 d_0^1 d_0^2 \ldots d_0^8\ (-1)^{\sum_{n=1}^{\infty} d^i_{-n}d^i_n +2\alpha'(\overline{p}^I)^2+2\alpha'(\tilde{p}^I)^2}$;

$\overline{W}^{\hat{i}}(k)$ is the vertex operator for the charged vector bosons $|\hat{i}\rangle_R \times |k_J\rangle_L$ <u>from a fermion line</u> [2] and Δ is calculated from the mass operator for a massive SS x Ramond sector sector compactified on a shifted internal momenta "lattice":

$$\Delta = \alpha'/2\Pi \int d^2z\, z^{\bar{N} + \alpha'(\bar{p}^I)^2}\, \bar{z}^{\tilde{N} + \alpha'(\tilde{p}^I)^2}\, |z|^{\frac{\alpha' p^2}{2} - 2} \tag{32}$$

$$\bar{N} = \sum_{n=1}^{\infty} A^i_{-n} A^i_n + 1/2\, n S_{-n} \gamma^- S_n \tag{33a}$$

$$\tilde{N} = \sum_{n=1}^{\infty} \tilde{A}^i_{-n} \tilde{A}^i_n + \quad n d^i_{-n} d^i_n \quad . \tag{33b}$$

From (29c), the allowed internal momenta for this sector cannot be zero and are given by

$$\sqrt{2\alpha'}\,\bar{p}^I = \sum_L (N + 1/2)^L \alpha_L{}^I$$

$$\sqrt{2\alpha'}\,\tilde{p}^I = \sum_L (N + 1/2)^L \alpha_L{}^I . \tag{33c}$$

Therefore, the left and right-moving degrees of freedom which contribute as intermediate states in (31) are given by another sector also satisfying (16-20), but where now the momenta in (17) and (20) are given by (33c) corresponding to a twisted boundary condition on these internal coordinates, and (21) is replaced by

$$d^i(\tau+\sigma) = \sum_{n=-\infty}^{\infty} d^i_n e^{-2in(\tau+\sigma)} \quad , \quad \{d^i_n, d^j_m\} = \delta_{n,-m}\, \delta^{ij} . \tag{34}$$

This is the Green-Schwarz superstring x the Ramond fermionic string compactified on the $(Z+1/2)^6$ lattice. The lowest mass state in this sector has $\alpha' m^2 = 3$, since in this sector the lowest momentum is $\sqrt{2\alpha'}\, p^I = (1/2, \dots, 1/2)$.

The projection $(-1)^{\bar{P}} = 1$ indicates the left-moving Ramond fermions are Majorana-Weyl. In fact, this projection does not contribute to (31) but it is required for the unitarity of one-loop amplitudes for external particles in the Ramond sector.

To check perturbative unitarity, we calculate the imaginary part of (27) and show that it is equal to products of tree amplitudes. Since (27) can be reexpressed as the sum of (30) and (31), the discontinuity can be calculated in the standard way using the Cutkosky prescription of replacing propagators with delta functions, and we find that the imaginary part of (27) is again an infinite sum of theta step fuctions, multipled by tree amplitudes, with thresholds given by the free particle spectrum of the compactified SS x NS and SS x Ramond sectors. Since we are forced by unitarity to include these two sectors, the ten-dimensional limit of our string is the type II superstring.

REFERENCES

1. R. Bluhm and L.Dolan, Phys.Lett. B169, 347 (1986).
2. R. Bluhm, L.Dolan and P. Goddard, Rockefeller preprint B_1/187.

A RELATION BETWEEN GAUGE GROUPS AND DIFFEOMORPHISM GROUPS

B. Julia.

ABSTRACT:
This article goes beyond the talk given two months ago in Bad Honnef. For the topics covered there and for an introduction to deformation theory we refer to [*]. Here we would like to formulate clearly the relation between the abelian gauge group of linearized relativity and the diffeomorphism group the deformation preserving the Poincaré group. It is best explained using the background field formalism. We shall also review a result of R. Wald on the essential uniqueness (and rigidity) of general relativity .

I. SYMMETRIES AND CONSERVATION LAWS.

I.1. Symmetries of equations of motion or of an action?

In [1] we focussed our attention on the deformations of Lie algebras and their representations but we mentioned that in practice the Gupta-Noether method of construction of non-linear couplings goes back and forth between transformation rules and actions or at least equations of motion. There is in fact a difference between action and equations of motion: the former imply the latter but firstly the converse is ambiguous if one does not put additional restrictions on the action and secondly it is impossible unless the Helmholtz consistency condition can be satisfied. Let us explain this:

a) <u>The inverse problem of variations</u>. A set of partial differential equations is to be interpreted as the variational derivatives of an action with respect to some field variables. But one may effect an arbitrary change of variables and take arbitrary independent combinations of the equations of motion. For one choice of variables and a choice of corresponding equations of motion the Helmholtz consistency condition is simply the symmetry of the second variational derivatives. The arbitrariness in these choices is partially eliminated by assuming a power series expansion for the theory in powers of the coupling constant(s).

On the other hand global symmetries of the action are scarcer than global symmetries of the equations of motion: in the absence of constraints they lead to canonical (symplectic) hamiltonian transformations and each global symmetry of the action leads via the Noether theorem to a conservation law.

b) The generalized Noether theorem. We refer to the proceedings of a previous conference in this series for the obvious proof but pretty formula [**].

Given an action invariant under some transformations parametrized by a closed p-form one can find a (d-p-1) current-form that is closed (conserved) on shell. The case of electrodynamics corresponds to p=0.

I.2 Doctor $T_{\mu\rho}$ and Misters $T_{\mu\rho}$.

We shall now review the classical folklore about the energy-momentum tensor and try and disentangle its various incarnations. First let us give some notation. Identities will be indicated by the equal sign = but equalities using equations of motion will be denoted by $\approx$ resp. $\simeq$ (resp. $\approx$) if they use matter resp. gravity (resp. both) equations of motion. Let us first discuss the currents in flat space.

a) The Noether current of translations. Given a local Lagrangian field theory that is invariant under (global) rigid translations one knows how to define a conserved current for each direction of translation and each coupled sector of the theory, for simplicity we will consider only one such sector. The conservation is valid MODULO the equations of motion of the fields involved. In a relativistically invariant theory these currents form a second rank tensor which is not symmetric unless the spin vanishes. Note that the Lagrangian density itself need only be invariant up to a total divergence. One can define the total "Canonical" energy-momentum tensor and write it as a closed vector valued (d-1) form :

$$dT_\rho + dt_\rho \approx 0 \quad .$$

where we have distinguished the matter part T_ρ and the gravity part t_ρ and where we considered gravity as an ordinary field theory. As the equation suggests both cannot be tensor-densities but t_ρ is a pseudo-tensor, we shall now handle both terms.

b) The Belinfante modification. It is also well known that when the Lagrangian depends only on the fields and their first derivatives one can add to the previous (canonical) energy-momentum tensor a total divergence that leaves it conserved ON-SHELL and cancels out in the integrated momentum [1]. Furthermore the new tensor is symmetric MODULO the equations of motion if the action is rigidly Lorentz invariant (in the sense of flat space theories). We notice that there is no other on-shell modification of the same form namely $\delta T^\mu{}_\rho = \partial_\nu A^{[\nu\mu]}{}_\rho$ that leads to a symmetric tensor, but that the symmetric part of the Canonical tensor is in general different from the Belinfante tensor even on-shell. Finally we arrive at

c) <u>The Rosenfeld tensor and the "consistency" condition</u>. It was independently discovered by Belinfante but we shall use its customary name [2] in order to avoid confusions. It denotes usually the right hand side of the Einstein equations, it corresponds to the matter sector. We must now distinguish two cases I and II, the first and second order formalisms with respectively the moving frame and the Lorentz connection or just the moving frame as the independent variables. In the first case the torsion is determined by the equation of motion for the connection and may differ from zero, whereas in the second case it vanishes automatically.

In case II the Rosenfeld tensor is in fact equal to the matter Belinfante tensor MODULO the equations of motion, in particular it is symmetrical in its two indices. The matter tensor satisfies a conservation identity namely it is COVARIANTLY conserved (modulo the MATTER equations of motion) provided the matter Lagrangian is a scalar density of the right weight under diffeomorphisms. Its symmetry is automatic because it is twice the derivative of the action with respect to the metric.

In case I the symmetry of the energy-momentum tensor follows from local Lorentz invariance and the matter equations of motion and the Rosenfeld tensor is defined to be the derivative with respect to the vierbein. We shall now consider a purely metric theory of gravity and shall return to the more general case below. Using differential forms we have:

$$D\,T_\rho^{ROS} \approx 0 \quad \text{with D the covariant derivative.}$$
$$\text{and } T_\rho^{ROS} \approx T_\rho^{CAN}$$

Let us note that the invariance under diffeomorphisms of the Einstein-Hilbert part of the Lagrangian leads correspondingly to an identity sometimes called "consistency" condition which holds fully OFF-SHELL. It is a consequence of the Bianchi identity for the curvature.

$$D_\mu R_\rho{}^\mu = 0$$

d) <u>More Noether and generalized currents</u>. Let us now consider a general one parameter subgroup of the diffeomorphism group of the form $dx^\mu = tX^\mu(x)$. Lorentz transformations lead to a Noether current which is equal to $x^\mu T^{\rho\nu} - x^\nu T^{\rho\mu}$ up to a total divergence ($T^{\mu\nu}$ is precisely the Belinfante tensor). Scale transformations coming from the diffeomorphism group are a special case of those usually considered because the dimension of a field is taken here to be geometrical i.e. equal to its number of covariant world indices minus the contravariant one (now mass terms are allowed!). Finally we can also compute the currents of the special conformal transformations, again we restrict ourselves to the geometrical dimensions and furthermore the conformal weights k, of a field are restricted. The case of a general 1 parameter subgroup is not much more difficult. We must be careful that the variation of the Lagrangian density is equal to the sum of 2 terms: a shift $tX^\mu\partial_\mu L$ and a local rescaling of weight $t\partial_\mu X^\mu$ so that it is still a total divergence. The conserved current associated to the vector field X is then

$$J^\mu = X^\nu T_\nu{}^\mu + \delta L/\delta\partial_\mu\varphi \, \partial_\rho X^\sigma \Sigma^\rho{}_\sigma \varphi.$$

where $T_\rho{}^\mu$ is the canonical Energy-momentum tensor and $\Sigma^\rho{}_\sigma$ is the spin operator on the field.

On the other hand one can derive the ON-SHELL conservation of generalized currents [**] when a Lagrangian field theory is invariant under symmetries parametrized by closed p-forms, the case of Yang-Mills gauge invariance is a combination of p=0 and p=1. This can be applied to the diffeomorphism group. We leave it as an exercise to show that there is a vector valued (d-2)-form whose divergence is equal to the Pseudo-energy-momentum tensor of gravity plus the Energy-momentum tensor of matter when there is any.

I.3 The Gupta program.

The (not entirely precise) idea [3] is to expand Einstein's equations in powers of the FIELDS and to reconstruct them order by order starting from their linearized form . One assumes that they come from an action then from the action to order n in the fields one computes the term of order n of the Belinfante tensor of the spin 2 field. One then adds it as a source on the right hand side of the equation from which one computes the next term in the action through the inverse problem of variations. This should be done ON-SHELL as we are writing the equations of motion but it is beset by ambiguities ! In fact I do not know of any complete realization of that program. One major criticism is that quite often use is made of the Rosenfeld tensor, but this presupposes knowledge of differential geometry as one couples flat space fields to a background metric, and one could just as well covariantize the spin 2 part directly. We shall discuss below (section II.1) various related approaches to this deformation of the spin 2 free motion and in particular the so-called Noether method. It deals with an action and is consequently OFF-SHELL. Essentially one starts from the free theory and adds to it a term of the form $gA\wedge J$ where the current is a Noether current for a rigid symmetry of the free action and A is the gauge field 1-form of the free theory. The method works beautifully as an expansion in powers of the coupling constant for the Yang-Mills action and has been extensively used in supergravity theory but the case of General Relativity [4 + references therein] remains more heuristic. Here we would like to clarify the group theoretical aspects in the case of Gravitation theory.

I.4 Group theoretical remark.

The Rosenfeld tensor density is NOT equal to the symmetric part of the canonical energy-momentum tensor so one may wonder why it is this tensor that is coupled to the spin 2 field. Let us recall the transformation law of the moving frame L^{α}:

$L=1+kE$ and $\delta E^{\alpha}{}_{\mu} = \partial_{\mu} c^{\alpha} + k c^{\rho} \partial_{\rho} E^{\alpha}{}_{\mu} + k \partial_{\mu} c^{\rho} E^{\alpha}{}_{\rho}$.

The first term reduces to an abelian gauge transformation (it is a symmetry of the free theory), the second term corresponds to a rigid translation but the third term is a spacetime dependent LINEAR transformation: where does it come from? It corresponds to the changes of coordinate axes and is responsible for the non-abelianeness of the diffeomorphism group. But one can express the "metricity" of the affine connection (the preservation of the metric) by the choice of gauges in which it is equal to the Lorentz connection. One may reduce the group GL(4,R) to the Lorentz group. We also know that the local anomalies of the Lorentz and diffeomorphism symmetries can be exchanged for the same reason [5].

Another fact is that the Rosenfeld tensor is equal to the Belinfante tensor on-shell. We are interested here in off-shell quantities because we would like to construct an action. Nevertheless the on-shell symmetry of the Belinfante tensor (if the theory is Lorentz invariant) and its conservation if the theory is invariant under translations suggest that one should consider the full Poincaré group.

We shall now study the possible deformations of the free spin 2 theory and try to follow the Yang-Mills analog as closely as possible. In particular we shall use the Cartan-Weyl moving frame theory, this allows the use of forms and increases the resemblance with the Yang-Mills case, it has the further advantage that it eliminates the field redefinition ambiguities.

II. DEFORMATIONS.

II.1 Various approaches.

Soon after the discovery of the Yang-Mills equations the concept of gauge symmetries emerged as a common feature of gravity and vector gauge theories [6]. The problem was formulated as the transformation of a rigid (global, or constant parameter) symmetry into a local (gauge) symmetry. Wyss and later the supergravity builders and Fronsdal shifted the emphasis and advocated the construction of interacting gauge theories from the free theories as the deformation of infinite Lie algebras. The uniqueness of such a deformation was however not completely clear especially in the case of General Relativity. Very recently R. Wald obtained some rather general results in this direction [7] and we shall report on them in part III.

On the other hand in the case of gravitation the choice of gauge group is not universally accepted, everybody agrees on the diffeomorphism subgroup but the choice of connection varies widely: affine linear, linear, affine Lorentz (=Poincaré) or Lorentz. E. Cartan introduced torsion in 1923, and this was rediscovered many times. The actions vary as a consequence, the most studied variants are the square of the Weyl tensor, the square of the Riemann tensor, and torsion squared terms. A new formalism called the group manifold approach plays down gauge invariance and leaves the group theoretical aspect of the couplings mysterious [8]. It was developed in order to simplify supergravities, and this is one of our motivations as well [***].

II.2 The Yang-Mills case (with group K).

a) The Background field formalism. It is instructive to relate it to the Noether method. So let us start from the Yang-Mills action and express the connection L as the sum of two terms $L=A+gB$. Note that we expand in the product of a field by the coupling constant g, it is done implicitly in the Noether method. This restricts the theory to be essentially independent of the coupling constant at the classical level as one can scale g away by rescaling the field. The curvature $R=dL+eL\wedge L$ is the sum of the background curvature $F=dA+eA\wedge A$ and the covariant curvature $G=g(DB+egB\wedge B)$ where the covariant differentiation DB is with respect to the background field A. The main observation is that the original gauge invariance admits two equivalent realizations the background gauge invariance:

(i) $\delta A=Da=da-e[a,A]$, $\delta B=-e[a,B]$ on the one hand and

(ii) $\delta' A=0$, $\delta' B=Db-eg[b,B]$ on the other.

One sees that (if $F=0$) the formula for dB'is loosely speaking the sum of an abelian gauge transformation law corresponding to its $g=0$ part plus a background transformation with $a=gb$; one speaks sometimes of "gauging" the rigid symmetry K corresponding to (i) with $Da=0$.

b) The Noether current. Let us start from the linearized theory around a general flat background ($F=0$). We have a canonical current given by the invariance under the group K of rigid transformations with $Da=0$. The Lagrangian is given by $L_{ym}° = +1/2Tr(G°\wedge *G°)$ (the Killing form is normalized to

-1) and the current 3-form J is proportional to $B_\wedge *G° - *G°_\wedge B$.
Lym° has an abelian gauge invariance :

$$\delta'°A = 0 \quad \text{and} \quad \delta'°B = Db \ .$$

One would like to recover the non-linear couplings of the B field. So one imposes the first correction (of order g) either of the action or of the transformation rules of the second type above, one fixes respectively $Lym^1 \sim Tr(B_\wedge J)$ or the g term in the transformation law (ii). Then one constructs iteratively further corrections in the transformation rules resp. the action so as to end up with an invariant theory.

c) The fully non-linear action. Let us for example ask for a transformation rule of form (ii). One finds that one must add a term $Lym^1 = +egTr(B_\wedge J)$ in order to cancel the db term in the variation, the careful reader will have noticed a factor 1/2 in front of the Noether current (B appears quadratically). Then one gets a new piece in the current to be plugged back into the action with the right coefficient and then the procedure stops ! Of course it does not seem very useful a priori to rederive the Yang-Mills action from the Yang-Mills action but the point is that one can forget about the background field A thereby breaking the invariance (i) and yet reobtain the non-linear theory by the above method provided one specifies the rigid symmetry to be preserved. Let us however notice that the choice of the canonical current (in the Yang-Mills case) would not work in the Gupta program, one would need to modify it by the divergence of a suitable antisymmetric tensor to get the correct equations. So let us stick to the off-shell deformation. One can formulate it more abstractly and say that one searches for a deformation of the abelian gauge algebra into any other gauge algebra that is INVARIANT under K a given rigid symmetry and under the Poincaré group. We would like to PROVE that this deformation is unique for a simple group K, this is more than so-called "rigidity" :we start from a highly non-rigid gauge symmetry and select a UNIQUE deformation. Recently Wald proved by assuming locality of the transformation rules on a given representation (a collection of n vector fields) that the only gauge algebras that can arise from such a deformation are the gauge algebras for any rigid Lie algebra of dimension n [7]. We hope to return to this later but for now we shall try to transpose this approach to the case of General Relativity. In particular we would like to understand HOW the abelian gauge group of the linear theory becomes the non-abelian (and non gauge) group of diffeomorphisms.

II.3 Gravitation.

Let us again start from the result and abstract from it a direct method of construction of the Einstein-Hilbert Lagrangian from the linearized theory namely the free spin 2 theory of Fierz and Pauli. The latter theory has an abelian gauge invariance under the shift of the spin 2 field $h_{\mu\nu}$ by a gauge like expression $\partial_\mu t_\nu + \partial_\nu t_\mu$. The symmetrization is related to the symmetry of $h_{\mu\nu}$. As noted in section I.4 it is clear that one should use a moving frame formalism to disentangle the rotation from the translation symmetries. This suppresses the symmetrization problem and allows a closer analogy with the Yang-Mills case although the analogy is NOT complete.

a) The Background field. We recall that Einstein's theory can be seen 1) as a Lagrangian theory of a metric field or 2) (Palatini) as a theory of a metric and an independent torsionless affine connection, or 3) as a theory with arbitrary metric and metric preserving affine connection, or 4) finally as a theory with independent (orthonormal) moving frame and Lorentz connection (Cartan Weyl). In the last formulation the rigid symmetries of the flat background form a Poincaré algebra of Killing vectors but the "gauge" symmetries are the diffeomorphisms plus the local Lorentz changes of frame. Let us again expand around a flat background one could probably also expand around de Sitter background (the following discussion should be extended to this case). So our fields are a soldering 1-form $X^\alpha = C^\alpha + kE^\alpha$ dual to the moving frame and a Lorentz connection 1-form $L^\alpha{}_\beta = A^\alpha{}_\beta + gB^\alpha{}_\beta$. Their variations can again be split in 2 different ways between background and perturbation fields. Firstly:

$$\text{(i)}\quad \delta C = dc + A.c - a.C = Dc - a.C \qquad \delta A = da - [a,A] = Da$$

$$\delta E' = t{*}DE' - [a,E'] \text{ and } \delta B' = t{*}DB' - [a,B'] + a^t.B' \ .$$

where D is the background A-covariant derivative, $E=E'.C$, and $B=B'.C$,($E'=E'^\alpha{}_\beta$ and $B'=B'^\alpha{}_{\beta\gamma}$), "." is the Lorentz contraction (in fact the matrix product) and * is the contraction of a vector and a 1-form. The last term in dB' is a contraction with the flattened index of the ex-1-form $B'^{\cdot\cdot}{}_\gamma$, E' and B' are Lorentz tensors and world scalars.
WARNING the transformation may look like a Poincaré gauge transformation but it is a diffeomorphism of flat space combined with a local Lorentz transformation (see [9] for example). What happened is that we made field dependent redefinitions of the parameters of the transformations (and of the fields) which modify the rules of computation of the Lie bracket. Precisely the new parameters are given in terms of the diffeomorphism parameters t^μ and those of the local Lorentz transformation $\Lambda^\alpha{}_\beta$ by

$c^{\alpha}=t*C^{\alpha}$ and $a^{\alpha}{}_{\beta}=\Lambda^{\alpha}{}_{\beta}+t*A^{\alpha}{}_{\beta}$.

The total curvature R is again the sum of 2 terms and we have R=F+gG . The Lagrangian reads (see for example[8])

$$Lgr = \epsilon_{\alpha\beta\gamma\delta}\, R^{\alpha\beta}{}_{\wedge}X^{\gamma}{}_{\wedge}X^{\delta} .$$

We shall use the convention that the ϵ tensor indices are omitted and contracted sequentially on the available Lorentz indices. Again $F= dA+A_{\wedge}A$ and $G= DB+gB_{\wedge}B$. Note that we have rescaled the fields so as to have e=1 for the field A. We also have a total torsion 2-form Z= S+kT where S= DC and $T= DE+g/kB_{\dot\wedge}C+gB_{\dot\wedge}E$. The second term in T could be combined with the first IF k and g did not vanish then by rescaling B and setting g=k one could interpret these 2 terms as the Poincaré covariant derivative of an adjoint multiplet (E,B). When we expand around a flat background with F=S=0 the Lagrangian density reads: (again . means the Lorentz contraction between adjacent indices)

$$Lgr = \epsilon\ [- 2gk\ C_{\dot\wedge}DB'_{\wedge}C_{\wedge}E + g^{2}\ B_{\dot\wedge}B_{\wedge}C_{\wedge}C$$

$$+ 2kg^{2}\ B_{\dot\wedge}B_{\wedge}C_{\wedge}E - k^{2}g\ C_{\dot\wedge}DB'_{\wedge}E_{\wedge}E + k^{2}g^{2}\ B'_{\dot\wedge}B'_{\wedge}E_{\wedge}E]$$

where B' is the purely Lorentzian tensor $C^{-1}*B$ again, D is the A-covariant derivative and the Lorentz contraction of C and B' is on the flattened index of B'. On the first line we have the quadratic part of the action. If k or g is equal to zero we see that there is no nonlinear term and even the free action becomes trivial so let assume that they do not vanish.

A second form of the transformation rules can be obtained as before ONLY if we take g and k of the same order, so we shall now take them to be equal without loss of generality. The second transformations preserve the background fields and can be expressed in terms of the variations (i) with parameters a and c but (ii) corresponds to $t= kC^{-1}.c$ and $\Lambda= ka-t*A$.

$$\text{(ii)}\qquad \delta'C = 0 \quad\text{and}\quad \delta'A = 0 \quad\text{but}$$
$$\delta'E = \delta C+k\delta E \quad\text{and}$$
$$\delta'B = \delta A+k\delta B .$$

The linearized theory has an abelian gauge invariance. Let us drop a factor k^{2} then:

$Lgr^{\circ}= \epsilon\ (B_{\dot\wedge}B_{\wedge}C_{\wedge}C -2C_{\dot\wedge}DB'_{\wedge}E_{\wedge}C)$ is invariant under:

$$\delta'^{\circ}E = Dc-a.C \qquad\text{and}\qquad \delta'^{\circ}B = Da$$

We see that B and E are abelian gauge fields but we note that the linearized Lorentz gauge invariance is broken as in

a sigma model and as expected for the moving frame. Let us take C=1 for background, then we see that the antisymmetric part of E can be gauged away.
The Lagrangian would not have been quadratic in the B and E fields if we had not set the cosmological constant to zero, of course had we expanded around De Sitter space the situation would have been different.

b) The Noether currents. We have assumed that the flat background field has a Poincaré algebra of Killing vectors. Then we have the corresponding conserved canonical currents of the quadratic part of the action. They are the canonical energy-momentum and spin currents. The coupling of the spin current works exactly as in the Yang-Mills case, but for the translation current it does not. The reason is that E' and B' transform as a Poincaré multiplet under (i) but not E and B which are the gauge fields. So we must compensate and change the Noether procedure. This will be postponed to a later publication. We would like now to present an apparently new symmetry of the Einstein-Cartan-Weyl... action in any dimension of spacetime.

c) The torsion symmetry. It turns out that given any vector field v^μ one can find a transformation of the Lorentz connection that preserves the action when combined with the following transformation law of the moving frame.

$$\delta E^\alpha = v \lrcorner T^\alpha$$

I plan to study its consequences more thoroughly elsewhere but it is relevant here because it is in 4 dimensions the difference between a diffeomorphism and the transformation considered in [10] and identified there as a diffeomorphism. The authors constructed the action for first order General Relativity using the Noether method for this symmetry. When the torsion does not vanish and it does not off-shell, this symmetry is genuinely different from a diffeomorphism. I discovered this little known article too late to explore its consequences here, it poses new problems for the deformation approach. Finally let us turn to a different discussion of General Relativity whose main originality (at least in my opinion) is the absence of the action but still with a choice of the manifold of field configurations on which the symmetries act. In a sense it lies inbetween the Noether approach and the purely abstract Lie theoretic approach that is so well suited to cohomological methods.

III. THE RIGIDITY OF THE DIFFEOMORPHISM GROUP.

We now turn to the result of [7]. If one assumes
1) that the gauge field equations obey "consistency" conditions to all orders in the perturbative fields,

2) that these conditions are local i.e. involve a (small) finite number of derivatives: they are assumed to be of the form
$\partial_\mu Eq^\mu . = a_\mu \, Eq^\mu . + b_\mu{}^\nu d_\nu Eq^\mu .$
(where the right hand side is the correction to the free consistency conditions and where only a_μ may involve the first derivative of the fields linearly but both a and b are also functions of the fields at one point),
3) that they generate a Lie algebra of infinitesimal symmetries of a corresponding (but not explicit) action with as many parameter functions as there are consistency conditions,
4) AND that the Poincaré covariance is preserved by the nonlinear couplings,
then the Lie algebra is either the abelian (linearized) spin 2 gauge invariance or else the algebra of vector fields on space-time.
There is a nice generalization to the case of several spin 2 fields. The whole discussion is purely algebraic, it involves only a nonlinear realization of a Lie algebra of local transformations. The original motivation was to preserve the "number" of degrees of freedom of the free equations of motion in the nonlinear deformation, so one required a nonlinear consistency equation for each linear one.
Let us make three remarks, the rigidity or absence of nontrivial deformations of the algebra of infinitesimal diffeomorphisms among "local" transformation Lie algebras was already discussed in [11]. The analysis of [7] implies rigidity but it implies also that there is no other deformation of the free algebra, at least no other deformation of the free algebra that PRESERVES the Poincaré invariance (or presumably the background Killing vectors if one could generalize [7] to De Sitter space). This means that rigid Poincaré invariance remains an automorphism of the gauge algebra. And finally it seems necessary to assume the existence of a Lie algebra of transformations, it is more restrictive than consistency conditions alone. Anyhow the justification of these conditions by the counting of degrees of freedom is more delicate for compact spaces because of linearization stability problems precisely when there are Killing vectors [121]. The space of solutions may admit a conical singularity and be unrelated to the space of solutions of the linearized equations. This is a problem for the theory in the large and goes beyond the local point of view adopted in this work, in particular compacity may break Poincaré invariance.

CONCLUSIONS.

We isolated the difference between the Yang-Mills gauge invariance and the diffeomorphism symmetry by the background field method and exhibited the implicit but crucial role of rigid Poincaré invariance at all stages of the deformation.

In string theory one must find the analog of this global symmetry in order to construct a geometric field theory of closed strings by the Noether approach. Finally many different cohomologies play a role in the Gupta-Noether program and their precise interplay deserves more study.

ACKNOWLEDGEMENTS. I am much obliged to the Mathematical S. R. Institute (Berkeley) for its generous hospitality (1984). I am also grateful to R. Wald for sending me his papers before publication.

REFERENCES.

[1] Belinfante F.J. Physica 7 (1940) 449.
[2] Rosenfeld L. Mém. Acad. Roy. de Belgique 18 (1938) 1 and Eddington A.S. Mathematical theory of relativity Cambridge Univ.Press (1937).
[3] Gupta S.N. Phys.Rev. 96 (1954) 1683, for a critical review of early work in this direction and a more precise approach see Fronsdal C. J. Math. Phys. 20 (1979) 2264.
[4] Deser S. Gen. Rel. and Grav. 1 (1970) 9.
The most useful application of this work was the construction of supergravities see for example
P. van Nieuwenhuizen's Physics Reports 68-4 (1981) 189.
[5] Bardeen W.A. and Zumino B. Nucl. Phys. B244 (1984) 421.
[6] Utiyama R. Phys. Rev. 101 (1956) 1597.
[7] Wald R. Phys. Rev. D33 (1986) 3613. and C. Cutler and R. Wald (Chicago preprint) A new type of gauge invariance for a collection of massless spin 2 fields.
[8] Regge T. Phys. Rep. 137 (1986) 31.
[9] von der Heyde P. Phys.Lett. 58A (1976) 141 and with F.W. Hehl and G.D. Kerlick Rev.Mod.Phys. 48 (1976) 393.
[10] Boulware D.G. Deser S. and Kay J. H. Physica 96A (1979) 141.
[11] Lichnerowicz A. C.R.Acad.Sc. Paris 281 (1975) 507 and Comment. Math. Helv. 39 (1976) 343. See also Fronsdal [3].
[12] Fischer A.E. and Marsden J.E. p.138 in General Relativity ed.S.W. Hawking and W. Israel Cambridge U.P.1979.
[*] B.Julia in the Proceedings of the Espoo Symposium on Topological and Geometrical Methods in Field Theory, World Scientific 1986 (to appear).
[**] B.Julia in the Proceedings of the Bad-Honnef conference 1980 (Baltimore preprint).
[***] B.Julia in Vertex operators in Mathematics and Physics J.Lepowski et al. 1984 p393 and in Recent developments in Quantum Field Theory J.Ambjorn et al. 1985 p215.

Recent Results in String Field Theory

A. Schwimmer
Laboratoire de Physique Théorique, ENS*
Paris, France
and
Department of Nuclear Physics, Weizmann Institute of Science
Rehovot, Israel

* Laboratoire propre du CNRS associé a l'Ecole Normale Superieure

The standard formulation of string theory gives rules for calculating Feynman-like, dual diagrams. A complete understanding of the symmetries as well as the calculation of non perturbative effects requires a Lagrangian reformulation leading to a String Field theory. The relation between such a formulation and the standard one is the string analog of the relation between the Lagrangian formulation of an ordinary field theory and the Feynman rules in the proper time formulation.

In the present contribution we will review the progress made in the Lagrangian formulation of the open bosonic string.

The effort of many groups[1] working independently led to a unique solution for the free Lagrangian along the lines first proposed by W. Siegel[2]. On the other hand, for the interacting Lagrangian there are two different proposals: the Kyoto-CERN[3] and the Lagrangian proposed by E. Witten[4]. We will discuss here, mainly the second proposal.

We start with a review of the basic building blocks and axioms in Witten's approach. The basic entity is the string functional $A(\dot{x}^{\mu}(\sigma), \phi(\sigma))$, $0 \leq \sigma \leq \pi$, where x_{μ} are the string coordinates and ϕ represents the bosonised Faddeev-Popov ghosts. The string functional is graded by ghost number+3/2, the physical sector being given by functionals with grade 1.

Using the fact that in the critical dimension the BRS operator Q is nilpotent, Q is used as an exterior derivative on the space of string functionals.

In order to define an integration and a product one singles out the middle of the string ($\sigma = \pi/2$), thus breaking explicitly reparametrization invariance at the first quantified level. The integral is defined then by:

$$\int A(x_\mu,\phi) \equiv \int Dx_\mu D\phi\, exp - \frac{3i\phi(\pi/2)}{2}\cdot A(x_\mu,\phi)\times \\ \times\prod_{\sigma=0}^{\pi/2}\delta(x_\mu(\sigma)-x_\mu(\pi-\sigma))\delta(\phi(\sigma)-\phi(\pi-\sigma)) \tag{1}$$

Geometrically, this represents the folding of the string on itself at $\sigma=\pi/2$ and it is non zero for functionals of grade 3. The explicit insertion of an operator depending on the ghost field takes care of the volume term in the ghost number anomaly[4].

The product $*$ of two functionals is defined to be non zero only when the right half of the first string overlaps with the left half of the second, in which case the result of the product is obtained by joining the non-overlapping pieces. In particular, the integrated product of three functionals is given by:

$$\int A_1 * A_2 * A_3 \equiv \int Dx_\mu^{(i)}D\phi^{(i)} exp\frac{3i\phi(\pi/2)}{2}\cdot\prod_{i=1}^{3}A_i(x_\mu^{(i)},\phi^{(i)})\times \\ \times\prod_{i,\mu}\prod_{\lambda=0}^{\pi/2}\delta(x_\mu^{(i+1)}(\lambda))\delta(\phi^{(i)}(\pi-\lambda)-\phi^{(i+1)}(\lambda)) \tag{2}$$

The operations defined this way fulfil the set of axioms[4]:

i) $Q^2=0$ ii) $\int QA=0$ iii) $A*(B*C)=(A*B)*C$ iv) $Q(A*B)=QA*B+(-1)^A A*QB$ v) $\int A*B=(-1)^{AB}\int B*A$.

where A in $(-1)^A$ is the grade of the functional.

The interaction Lagrangian is constructed then, in analogy with the three dimensional "gauge invariant mass term", as:

$$I=\int(A*QA+\frac{2}{3}A*A*A) \tag{3}$$

Using the axioms, it is easy to show that I is invariant under the gauge transformations:

$$\delta A = Q\varepsilon + A * \varepsilon - \varepsilon * A \tag{4}$$

The gauge transformations form a group under the composition law:

$$\varepsilon_{12} = \varepsilon_1 + \varepsilon_2 + \varepsilon_1 * \varepsilon_2 - \varepsilon_2 \star \varepsilon_1 \tag{5}$$

The guiding principle behind the construction of I was to insure the gauge invariance (4). This invariance should incorporate the various local invariances present in the first quantized formulation. It is by no means clear that the postulated Lagrangian reproduces the results of the standard formulation. The discussion of this issue is the main topic in this review.

We start with a discussion of the free spectrum: the free equation of motion is:

$$QA = 0 \text{ with gauge invariance } A \rightarrow A + Q\varepsilon \tag{6}$$

Therefore finding the free spectrum is really the problem of finding the cohomology of Q. This problem was solved in Refs. 2 and 5, which show that the solutions of (6) are in one to one correspondence with the physical states of the Veneziano Model.

We discuss now the vertex function[6,7]. The Lagrangian (3) gives a unique expression (2) for the interaction vertex as an integral over a triple product. This prescription looks geometrically rather different from the one appearing in the light cone formulation[8,9] (Fig. 1). On the other hand, the couplings given by (2) should coincide with the standard ones (reproduced correctly by Fig. 1a) just on physical states. In order to make the comparison possible one needs to rewrite (2) in an oscillator representation, i.e. one needs to find a vertex operator $V(a^i, b^i)$ such that the couplings are given by the matrix elements:

$$\underset{1,2,3}{<} 0|V(a^i, b^i)|1>|2>|3> \tag{7}$$

where $|i>$, a^i, b^i, i=1,2,3 are string states and operators (a - coordinates, b - ghosts) respectively. The technique for finding (7) once an overlap of type (2) is given was formulated by Mandelstam[8]. The basic idea of the procedure is to isolate (7) from a time evolution matrix element calculated in the momentum representation. The matrix element is calculated in the path integral representation, the time evolution producing strips attached to the vertex (Fig. 2). The special feature of this domain is the identification of the segments OAB and OA′B′ as a consequence of the special form of the Witten product. As explained by Mandelstam, $V(a^i, b^i)$ is obtained from the asymptotic expansion of the Neumann function for the domain of Fig. 2. To find the Neumann function one maps Fig. 2 into a domain (e.g. the upper half plane) for which the Neumann function is known. The mapping is done in two stages. First we eliminate the double sheeted nature and identification in Fig. 2 through the mapping:

$$\chi = \rho^{2/3} \tag{8}$$

where $\rho = i\varsigma - \sigma$ is the variable in Fig. 2 and χ the one in Fig. 3 which is the new domain. The segments OAB and OA′B′ are now overlapping and the identification is replaced by the requirement of continuity across OB. The mapping (8) defines a second order differential

$$d\rho^2 = \frac{9}{4}\chi d\chi^2 \tag{9}$$

the linear zero of which has an invariant meaning and characterizes the world sheets obtained in Witten's formulation[10].

After the mapping to Fig. 3, the further mapping to a half plane (with variable z) becomes a standard problem in comformal mappings. The combination of the two mappings gives:

$$z(\varsigma) = \frac{1}{2} + \frac{i\sqrt{3}}{4}\left(\varsigma + \frac{1}{\varsigma}\right) + \frac{i\sqrt{3}}{4}\left(\varsigma - \frac{1}{\varsigma}\right)\left[Y_+(\varsigma)e^{-\pi i/3} + Y_-(\varsigma)e^{\pi i/3}\right] \tag{10}$$

where ς=exp(-iρ) and

$$Y_+(\varsigma) = Y_-(-\varsigma) = \left(\frac{1-\varsigma}{1+\varsigma}\right)^{1/3} \tag{11}$$

A lengthy calculation gives finally for the vertex[6]:

$$\begin{aligned} V = exp \Bigg\{ & \frac{1}{2}\sum_{m,n=1}^{\infty} (a_r^m \cdot a_s^n + b_r^m b_s^n)\, N_{mn,rs} + \\ & + \sum_{\substack{n=1 \\ odd}}^{\infty} \left[a_r^n \cdot \sqrt{2}\,(P_{r+1} - P_{r+2}) + b_r^n\,(G_{r+1} - G_{r+2})\right] N_{n0,11} \\ & + \sum_{\substack{n=2 \\ even}}^{\infty} \left[a_r^n \cdot \sqrt{2} P_r + b_r^n \left(G_r N_{n0,11} + (-1)^{n/2}\frac{y_n - 2}{2n}\right)\right]\Bigg\} \end{aligned} \tag{12}$$

r,s = 1,2,3, where:

$$N_{mn,rr} = (-)^{\frac{m+n}{2}}\left\{\frac{(-)^m + (-)^n}{3(m+n)}(2W_{mn} - y_m y_n) - \frac{\delta mn}{3m}\right\} \tag{13a}$$

$$N_{mn,r\,r+1} = -(-)^{\frac{m+n}{2}}\left\{\frac{(-)^m + (-)^n}{6(m+n)} + (-\sqrt{3})\frac{(-)^m - (-)^n}{6(m+n)}\right\}(2W_{mn} - y_m y_n) - \frac{(-)^n}{3m}\delta_{mn} \tag{13b}$$

$$N_{n0,11} = \begin{cases} (-)^{n/2}\frac{y_n}{n} & \text{n even} \\ (-)^{\frac{n+1}{2}}\frac{1}{\sqrt{3}}\frac{y_n}{n} & \text{n odd} \end{cases} \tag{13c}$$

$$W_{mn} = \begin{cases} -\frac{3}{2}\frac{m(n+1)y_m y_{n+1} - n(m+1)y_{m+1}y_n}{m-n} & m \neq n \\ \sum_{k=0}^{n}(-)^{k+n}y_k^2 & m = n \end{cases} \tag{13d}$$

and y_n is defined by:

$$\left(\frac{1-x}{1+x}\right)^{1/3} = \sum_{n=0}^{\infty} y_n x^n \tag{14}$$

The momenta P_r and the ghost numbers G_r are constrained by

$$\sum_1^3 P_r = 0 \qquad \sum_1^3 G_r = -\frac{3}{2} \tag{15}$$

Using (12) one can calculate the couplings of the low lying physical states, the results agreeing with the couplings in the standard formulation. A general proof for the equality of all the physical state couplings can be formulated[11]. The essence of the proof is that once a mapping between a standard formulation (e.g. the light cone) and the Witten configuration exists, the differences between the various vertices involve only exponentiated Virasoro operators of positive index which give one on physical states.

The physical states couplings being consistently reproduced, one is led to the study of Feynman diagrams in the String Field theory and their comparison with the ones appearing in the standard treatment. We mention first a remarkable result obtained in Ref. 12 concerning the integration region in the space of parameters. In the field theory one integrates over a set of parameters which are the proper times in the Schwinger formula for the propagators. On the other hand in the first quantized formulation one has for each topology of the world sheet a set of modular parameters. Using a theorem proven by J. Harer[13], Ref. 12 shows that there is a one-to-one mapping between the two parameter spaces mentioned above, i.e. Witten's string field theory provides a natural triangulation of the modular parameter space. This result uses the specific shape of the interaction vertex and it is not true, e.g. for the light cone interaction vertex. The mapping of the integration regions is just a necessary condition for reproducing the amplitudes. One needs to reproduce also the integrands and in particular, the measure on parameter space. This comparison was done by Giddings[14] for the four point tree diagram and extended in Ref. 10 for all N-point tree amplitudes. Starting with the Lagrangian in the Siegel gauge ($b^0A=0$), the propagators contain an insertion of a line integral over the b-ghost field. This insertion reproduces the correct measure for the choice of modular parameters given by the proper times.

The explicit ghost insertions in the Witten vertex reproduce the correct ghost number violating Lagrangian and give a boundary term which changes the ghost number of the external states from $-\frac{1}{2}$ in the Siegel gauge to $+1$ in the first quantized formalism. This is the general mechanism which assures that the tree amplitudes of the first quantized formulation are correctly reproduced by Witten's Lagrangian.

Calculations of the loop amplitudes in the Lagrangian formulation were prevented until recently by the "ghost of ghost" problem: in the Siegel gauge the free part of the Faddeev-Popov ghost string functional ("Siegel ghosts" in the following) Lagrangian has a gauge invariance not shared by the interactive part. This prevents a consistent gauge fixing necessary to define the propagators for the Siegel-ghosts. As a consequence loops in which the Siegel ghosts circulate could not be calculated.

The "ghost of ghost" problem was solved recently by Thorn[15] following a different path and therefore the calculation of loops is now, in principle, possible. Of particular interest is the calculation of the Freund-Rivers diagram[16] in which the closed string may appear as a pole. The integration region responsible for the pole is present[12] but it is essential to check that e.g. the pole is not cancelled by the contribution of the Siegel ghosts. The presence of the closed string in the open string theory would raise very interesting questions concerning the consistency of the theory and in particular the necessity for an independent closed string Lagrangian. If indeed the closed string pole appears, it suggests that the open string theory has a higher symmetry than expected including reparametrization (Einstein) invariance in real space-time.

Witten's inspired guess for the string field theory of the open bosonic strings reproduces successfully the first quantized formulation and constitutes a hopeful starting point for the Lagrangian formulation of the physically relevant heterotic string theories.

Acknowledgements

Very useful discussions on the topics contained in this contribution with D. Amati, E. Cremmer, J.L. Gervais, A. Jevicki, E. Martinec, S. Samuel, Ch. Thorn and G. Veneziano are gratefully acknowledged.

References

1. For a rather complete list of references see T. Banks: Invited lecture series at the Spring School on Supersymmetry, Supergravity and Superstrings, SLAC-PUB-3996, June 1986.
2. W. Siegel, Phys. Lett. **149B**, 157 (1984), Phys. Lett. **151B**, 391 (1985).
3. H. Hata, K. Itoh, T. Kugo, H. Kumimoto and K. Ozawa, Phys. Lett. **172B**, 186 (1986).

   A, Neveu and P. West, Phys. Lett.**168B**, 192 (1986).
4. E. Witten, Nucl. Phys. **B268**, 253 (1986).
5. D. Freeman and D. Olive, Imperial College preprint (1986).

   Ch. Thorn, Univ. of Florida, Gainesville preprint (1986).

   M. Spiegelglass, IAS Princeton preprint (1986).

   I. Frenkel, L. Garland, G. Zuckerman, Yale University Mathematics preprint (1986).
6. E. Cremmer, A. Schwimmer and Ch. Thorn, Phys. Lett. B (in press).

   Ch. Thorn, Contribution to the 23rd Int. Conf. on High Energy Physics, Berkeley, CA (1986).
7. D. Gross and A. Jevicki, Princeton preprint (1986).

   S. Samuel, CERNTH 4365 (1986).
8. S. Mandelstam, Nucl. Phys. **B64**, 205 (1973).
9. M. Kaku and K. Kikkawa, pr **D10**, 1110, 1823 (1974).

E. Cremmer and J.L. Gervais, Nucl. Phys. **B76**, 209 (1974).

10. S. Giddings and E. Martinec, Princeton preprint (1986).

11. E. Cremmer, A. Schwimmer and Ch. Thorn (unpublished).

    S. Samuel, CERNTH preprint (1986).

12. S.Giddings, E. Martinec and E. Witten, Princeton preprint (1986).

13. J. Harer, Maryland Math. preprint (1984).

14. S. Giddings, Princeton preprint (1986).

15. Ch. Thorn, private communication.

16. P.G.O. Freund and R.J. Rivers, Phys. Lett. **29B**, 510 (1969).

Figure Captions

Fig. 1: The light cone (a) and the Witten (b) three open string interactions.

Fig. 2: The interacting string diagram corresponding to Fig. 1b. The diagram is drawn on two Riemann sheets with cut on the positive real axis. Dotted lines lie on the second sheet.

Fig. 3: The image of Fig. 2 in the χ plane which displays the three strings on an equal footing.

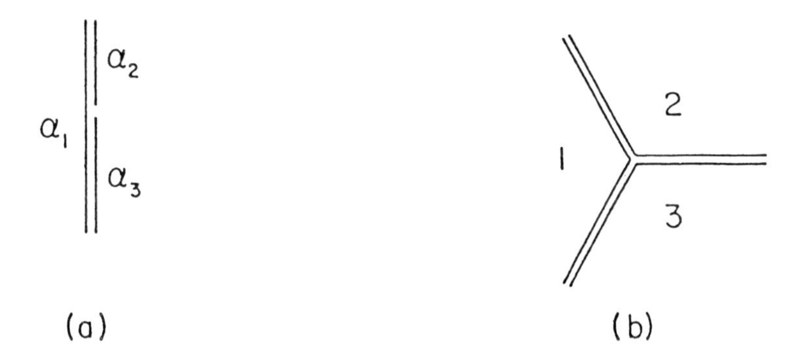

Figure 1

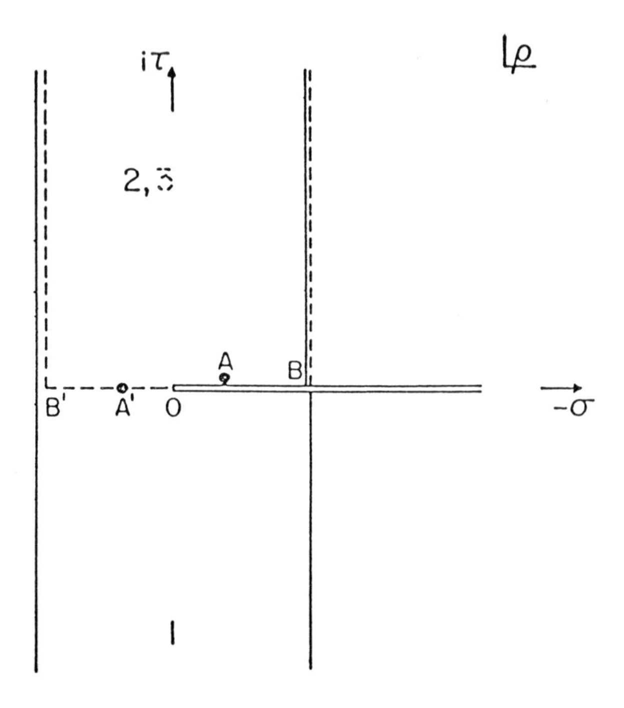

Figure 2

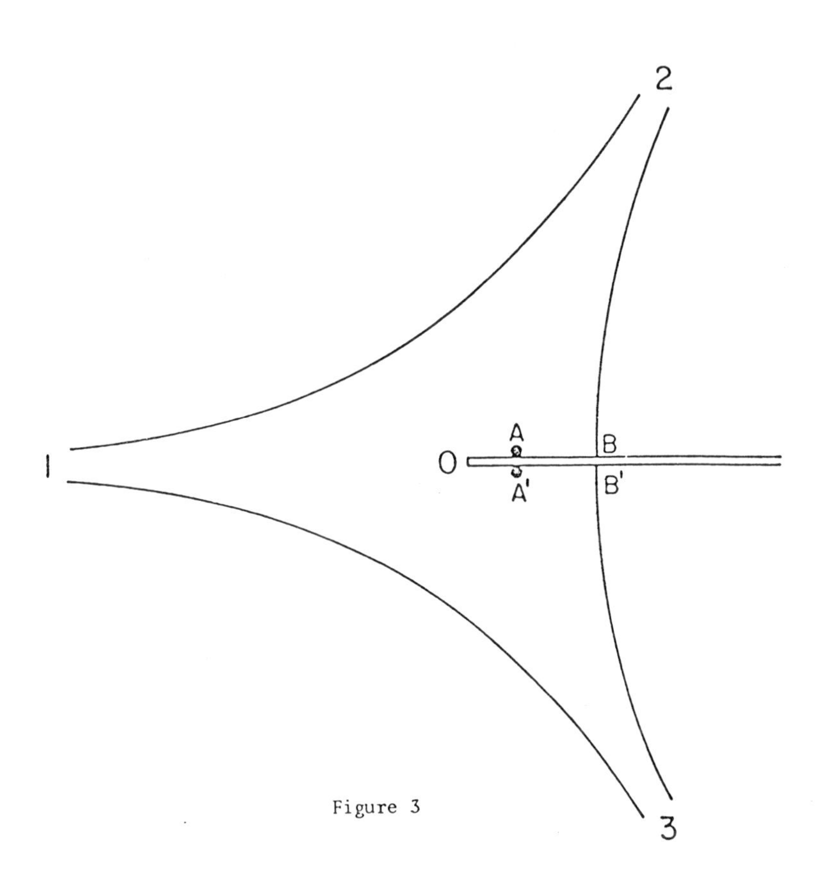

Figure 3